An Introduction to
Sustainable Development

This fourth edition has been comprehensively rewritten and updated to provide a concise, well illustrated and accessible introduction to the characteristics, challenges and opportunities of sustainable development with particular reference to developing countries. The contested nature of sustainable development is explored through a detailed consideration of changing ideas and practices within environmentalism and development thinking. The text identifies the different actors involved (from institutions of global governance through to community based organisations), the policies and mechanisms through which sustainable development is being sought, and considers the outcomes for particular groups and environments in both rural and urban contexts.

This edition places stronger emphasis on the global challenges of sustainable development with an understanding of interlinked crises in climate, energy, economy, poverty and social injustice. It explores how these issues are leading to deep questioning of what sustainable development is, what it should be, and how sustainable development policies and mechanisms are being reconsidered. The book gives new consideration to the challenge of achieving lower carbon growth, climate adaptation, and the implications on sustainable development of rapidly expanding economies, including China and India. It contains greater discussion of how civil society movements influence outcomes of international climate policy, as well as technological developments in energy and agriculture. The text also contains a substantially expanded discussion of how poverty remains central to sustainable development challenges, as revealed through the Millennium Ecosystem Assessment and Millennium Development Goals.

This invaluable text retains the core message that sustainable development has become central to debates about environment and development. Containing a substantial number of new boxed case studies, learning outcomes, chapter summaries, discussion questions, further reading and websites, this text provides an essential introduction for students.

Jennifer A. Elliott is Principal Lecturer in Geography at the University of Brighton.

Routledge Perspectives on Development

Edited by: Professor Tony Binns, *University of Otago*

The *Perspectives on Development* series will provide an invaluable, up to date and refreshing approach to key development issues for academics and students working in the field of development, in disciplines such as anthropology, economics, geography, international relations, politics and sociology. The series will also be of particular interest to those working in interdisciplinary fields, such as area studies (African, Asian and Latin American Studies), development studies, rural and urban studies, travel and tourism.

If you would like to submit a book proposal for the series, please contact Tony Binns on j.a.binns@geography.otago.ac.nz

Published:

Third World Cities, 2nd edition
David W. Drakakis-Smith

Rural–Urban Interactions in the Developing World
Kenneth Lynch

Environmental Management and Development
Chris Barrow

Tourism and Development
Richard Sharpley and David J. Telfer

Southeast Asian Development
Andrew McGregor

Population and Development
W.T.S. Gould

Postcolonialism and Development
Cheryl McEwan

Conflict and Development
Andrew Williams and Roger MacGinty

Disaster and Development
Andrew Collins

Non-Governmental Organisations and Development
David Lewis and Nazneen Kanji

Cities and Development
Jo Beall

Gender and Development, 2nd edition
Janet Henshall Momsen

Economics and Development Studies
Michael Tribe, Frederick Nixon and Andrew Sumner

Water Resources and Development
Clive Agnew and Philip Woodhouse

Theories and Practices of Development, 2nd Edition
Katie Willis

Food and Development
E. M. Young

An Introduction to Sustainable Development, 4th edition
Jennifer Elliott

Forthcoming:

Global Finance and Development
David Hudson

Natural Resource Extraction and Development
Roy Maconachie and Gavin M. Hilson

Politics and Development
Heather Marquette and Tom Hewitt

Children, Youth and Development, 2nd Edition
Nicola Ansell

Climate Change and Development
Thomas Tanner and Leo Horn-Phathanothai

Religion and Development
Emma Tomalin

Development Organizations
Rebecca Shaaf

Latin American Development
Julie Cupples

An Introduction to Sustainable Development

Fourth edition

Jennifer A. Elliott

Routledge
Taylor & Francis Group

LONDON AND NEW YORK

First edition published
by Routledge 1994

Second edition published
by Routledge 1999

Third edition published
by Routledge 2006

This edition published 2013
by Routledge
2 Park Square, Milton Park, Abingdon, Oxon OX14 4RN

Simultaneously published in the USA and Canada
by Routledge
711 Third Avenue, New York, NY 10017

Routledge is an imprint of the Taylor & Francis Group, an informa business

© 1994, 1999, 2006, 2013 Jennifer A. Elliott

The right of Jennifer A. Elliott to be identified as author of this work
has been asserted by her in accordance with sections 77 and 78
of the Copyright, Designs and Patents Act 1988.

British Library Cataloguing in Publication Data
A catalogue record for this book is available from the British Library

Library of Congress Cataloging in Publication Data
A catalog record for this book has been requested

ISBN: 978-0-415-59072-3 (hbk)
ISBN: 978-0-415-59073-0 (pbk)
ISBN: 978-0-203-84417-5 (ebk)

Typeset in Times New Roman and Franklin Gothic by
Florence Production Ltd, Stoodleigh, Devon

MIX
Paper from
responsible sources
FSC
www.fsc.org   FSC® C004839

Printed and bound in Great Britain by the MPG Books Group

In memory of my dad

Contents

Plates

Figures

Tables

Boxes

Acknowledgements

I would like to thank my colleagues and friends for continuing to be so supportive and patient with me in relation to writing this new edition. My biggest thanks are to my family and to Hils who will be the most pleased that it is at last completed.

Introduction

This book is concerned with the continued challenges and
opportunities of finding more sustainable patterns and processes of
development within the international community for the future. Since
the publication of the first edition of this text in 1994, much has been
learnt regarding the principles of sustainable development, the
characteristics of policies, mechanisms and projects that appear to be
more sustainable, and the assessment and monitoring of environment
and development outcomes that are central to sustainable
development. Whilst the idea of sustainable development was
relatively new in 1994, it is now suggested to have 'come of age'
and forms a staple part of most debates about environment and
development (Redclift, 2005; Adams, 2009). The pursuit of
sustainable development is now stated as a principal policy goal of
many of the major institutions of the world including the United
Nations, the World Bank and the World Trade Organisation. In 2000,
the international community committed to achieving eight
'Millennium Development Goals' by 2015. One of these goals refers
explicitly to sustainable development, but all are central in that they
commit to better and more equitable outcomes in arenas such as
health, gender, housing and sanitation that directly affect poorer
groups. Poverty is a major cause and effect of global environmental
problems and addressing poverty and inequality are long-standing
and central concerns of sustainable development. The global
challenge of sustainable development has been confirmed in recent
years by the interlinked 'crises' of climate, economic recession and

rising food, fuel and commodity prices that impact hardest and first on the poorest people in societies. Finding ways to address and prevent these crises requires interconnected and interdisciplinary thinking that is also at the core of sustainable development.

Evidently, the context in which sustainable development is currently being pursued is significantly different to that in the 1990s. An increasingly globalised world has brought new challenges and opportunities for the environment and for development. New actors (such as transnational corporations and civil society organisations) and new technologies (particularly in computing, information and communication) now shape outcomes in resource development and management to a much greater extent than previously. However, the closer and deeper integration of people and places around the globe brings new risks as well as opportunities. Farmers, for example, may be able to access new and wider markets for their produce but have less direct control over decisions regarding what to grow and when to sell, to whom. They become increasingly vulnerable to changes in price and consumers' tastes set at great distances away.

Climate scientists have also now established the human causes of climate change. Yet existing patterns of economic development remain closely associated with increased energy demands and rising fossil fuel use. Moving towards lower carbon patterns and processes of development is a challenge for individuals, business and industry, governments and international organisations globally. However, contributions to processes of climate change, the experiences of its impacts and capacity to cope with change already occurring are not evenly distributed; within current societies, across different countries or between generations. This is one illustration of the complex interconnections between environmental resources and the functions and services they provide for human wellbeing and development. It also highlights that the challenges and opportunities of sustainable development are context specific, that is, they lie in the interconnections of factors of the natural and human environment in particular places and points in time ensuring that there is no simple or single 'route' to sustainable development.

Economic growth in the past two decades has delivered vast improvements in human well-being including moving over 400 million people out of poverty. Many of the fastest rates of economic growth currently are now in countries of the Global South. Brazil, Russia, India and China (the 'BRIC' economies) for example, are now responsible for a significant proportion of world exports and constitute a powerful group within international trade negotiations. However, recent economic success has been very unequal across and within countries; low-income countries (and particularly within the African continent) remain largely peripheral in terms of world trade and foreign investment, for example. Income inequality is also increasing worldwide. Differences in wealth and income are seen through the text to be important factors in explaining a range of spatial patterns of 'unsustainable development'. However, poverty has many dimensions beyond material wealth including the opportunity to participate in decisions that affect one's immediate environment and to feel valued within local communities. 'Making globalisation work better for the poor' is understood as integral to many of the challenges of sustainable development; as a human rights issue, as a moral concern, for peace and security and economic development in the future.

The primary foci of the book are the challenges and opportunities for sustainable development in the less economically developed regions of the world. Fundamentally, this is because it is here that the majority of the world's poor reside. This is not to suggest that sustainable development is mostly a problem for the poor. Indeed, most pollution (including carbon emissions) and resource consumption are a result of affluence, not poverty. There are deep challenges for sustainable development associated with changing consumption patterns in particular in the Global North. These are identified in the text, but a full consideration is beyond the scope of this particular book. The prospects of sustainable development in any one location are also very evidently shaped in part by forces and decision-making which are often situated at great distances away. It is impossible therefore to consider the developing world in isolation from the wider global community. Furthermore, some of the most innovative responses to the challenges of sustainable development are now being seen in countries within these developing regions. However, there are also particular and distinct issues of sustainable development in the developing world. For example, these regions encompass many of the world's 'fragile lands', such as the major arid and semi-arid zones and forest ecosystems. In these places, bio-physical factors in combination with some of the lowest levels of human development worldwide can make them particularly susceptible to degradation (including through climate change) and make recovery from natural and economic shocks and disturbance difficult.

Urbanisation is also occurring most rapidly in developing regions, particularly within Asia and Africa. Cities can present a number of advantages for sustainable development. For example the density of population can enable infrastructure such as public transport and waste disposal to be provided more efficiently and cost-effectively. Such services have major environmental and health benefits. However, the numbers of people in cities of the developing world living in slum conditions where basic services in water supply and sanitation (and in housing) are entirely lacking or severely compromised is rising not falling.

A key aim of the book is to highlight the progress that has been made towards establishing new patterns and processes of development which are considered more sustainable; in terms of the demands they make on the physical, ecological and cultural resources of the globe, and the characteristics of technology, societal organisation and economic production which underpin them. Understanding the characteristics of successful sustainable development projects will be essential for meeting the worldwide ongoing and evolving challenges of balancing present needs against those of the future. However, not only has the world changed since the publication of the first edition of this text, so too has how the challenges of sustainable development are understood. There are now new debates concerning what sustainable development *is* and *should* be and comprehensive critiques of the policies and mechanisms that have been used towards meeting the challenges. For example, past policies based on the privatisation of environmental resources such as in water supply and in forest conservation are being challenged (often through the action and resistance of people at the local level). There is also much rethinking particularly in the context of global economic recession of 'conventional' ideas about the links between economic growth and human well-being.

As the term 'sustainable development' reaches further into popular consciences worldwide, as more institutions recognise sustainability as a major policy goal, and at a time of a heightened sense of urgency in many arenas of environment and development, there is a continued need to reflect critically on what is trying to be achieved, whose interests and values may be dominant and what the costs and benefits of particular interventions, that is, policy responses and management decisions, are for particular people and local environments. The origins of the notion of sustainable development, its varied 'meanings' and the contribution of different disciplines are traced in Chapter 1 within an analysis of thinking and practice in development theory and in environmentalism. Whilst the interdependence of future environment and development ends has been embraced in both literatures, it is seen that substantial debate and contestation characterise both the theory and practice of sustainable development. The historical overview presented confirms that the context within which environment and development objectives are being pursued is changing rapidly, requiring continuous re-evaluation of the meaning of sustainable development as presented within particular schools of thinking and major international summits, for example. The chapter also identifies a number of ways in which 'mainstream' ideas and approaches to the global agenda of sustainable development are being challenged, including through better understanding of the environment and development concerns of the poor, in response to failures of development on the ground and through change led by citizens, practitioners and academics from within the developing world.

In Chapter 2, the contemporary global challenges of sustainable development are considered in some detail. It is seen that development continues to depend heavily on natural resources for an increasing number of functions but that inequality in access to resources has also been a persistent and entrenched feature of past development patterns and processes. Such inequality is seen to underpin substantial human insecurity, conflict, ill-health and premature death as well as resource degradation, confirming that development is not meeting the needs of current generations. In addition, the increasing global-scale impacts of human activities such as through climate change raise very starkly the question of current development compromising the opportunities of future generations.

In Chapter 3, a range of actions which are being taken at a variety of levels by some of the core institutions in development towards ensuring more sustainable processes and patterns are identified. Importantly, development (and environmental management) is no longer something undertaken principally by governments: rather many different kinds of formal organisation (including transnational corporations, non-governmental organisations and international financial institutions such as the World Bank) and less formal arrangements such as within communities and even households, influence environmental actions and outcomes. A key aim of the chapter is to consider how these actors in development are changing what they do, but also how they are working in new ways, together, to address the integrated challenges of sustainable development. The chapter also considers a number of 'cross-cutting' issues of trade, aid and debt, that illustrate the ways in which people and places across the globe are interconnected but also how these issues operate to shape the capacities of particular actors in development.

In Chapters 4 and 5, the particular challenges and opportunities of sustainable development in the developing world are considered in rural and urban contexts. It is quickly seen that the two sectors are not distinct and that the environment and development concerns therein are often interrelated. Indeed, one of the limitations of past development policies has been their tendency to consider rural and urban areas separately, and there is now better understanding of the complex and multidirectional linkages between the two contexts that shape landscapes and livelihoods. However, important differences are also seen, including in terms of the immediate environmental problems, the options for securing income and livelihood, the hazards and sources of instability of living and the specific opportunities for action. The principles which are seen to be now guiding more sustainable development interventions in practice, however, are regularly common to both rural and urban settings. Addressing the welfare needs of the poorest groups, building participatory systems of research and development and aspects of local governance are identified as being essential to achieving the goals of development and conservation in both contexts.

In Chapter 6, progress made towards sustainable development is considered through the expanding field of sustainable development indicators and appraisal. The development of indicators of sustainable development is an important part of reflecting on what is trying to be achieved (the vision of 'sustainable development') as well as for measuring the outcomes of specific policies and interventions and monitoring progress towards those goals (i.e. communicating 'how far' from sustainable development we are). The processes through which indictors are developed and assessments are made are also proving significant in developing the conversations and debate amongst different interest groups and empowering individuals and groups of people to take action on sustainable development. The evidence that a number of these 'alternative' measures of 'human progress' are revealing is now contributing to wider reflections on how development has been pursued. The final section of the text considers whether a 'common future' for sustainable development can currently be identified on the basis of reflections on the substantive chapters of the book.

Since the publication of the first edition of this text, there have been many reminders of the very direct relationship between human society and the resources and environmental processes of the globe. These have included tsunamis in Thailand, Sri Lanka and Japan (originating in earthquake activity) and war in Iraq and Afghanistan (that cannot be divorced entirely from the geography of oil resources). All have led to the loss of thousands of lives and removed basic development opportunities for many more. Through this book, the challenges of sustainable development will certainly be seen to encompass better scientific understanding of environmental processes and more cooperative and democratic international collaborations. But they will also be seen to include for example the accountability of industry to stakeholders and the power of all individuals to participate in the decisions that shape the opportunities for their own development.

① What is sustainable development?

Learning outcomes

At the end of this chapter you should be able to:

- Understand why sustainable development is a contested concept
- Understand that sustainable development requires thinking holistically on linked processes of environmental, economic and social change and with regard to the future
- Be aware of the historical origins of the idea of sustainable development
- Appreciate that sustainable development is considered an important challenge by international institutions and governments worldwide
- Identify the principal strengths and weaknesses of past approaches to development and environmental management

Key concepts

Sustainable development; development theory; globalisation; neo-liberalism; environmentalism; ecological modernisation; environmental justice.

Introduction

In 1984, the United Nations (UN) established an independent group of 22 people drawn from member states of both the developing and developed worlds, and charged them with identifying long-term environmental strategies for the international community. The report of the World Commission on Environment and Development entitled *Our Common Future* (WCED, 1987) is widely considered to have been key in putting sustainable development firmly into the political arena of international development thinking. It used the term 'sustainable development' extensively and defined it as 'Development that meets the needs of the present without compromising the ability of future generations to meet their own needs' (p. 43). The report has been translated into more than

24 languages (Finger, 1994) and its definition of the term continues to be that which is most widely used and cited. For the first time, the Commission had considered environmental concerns arising through development processes from an economic, social and political perspective rather than solely from a science base as in previous studies. Their recommendations focused on integrating development strategies and environmental policies and global partnerships to meet the interdependent environmental concerns and development opportunities North and South.

The work of the commission was undertaken as the basis for a UN conference on Environment and Development to be held five years later. The 'Earth Summit' in Rio de Janeiro, Brazil in 1992 was, at the time, the largest ever international conference held. It was also the first time heads of state had gathered to consider the environment. One hundred and sixteen heads of state or governments and over 8,000 delegates attended. A further 3,000 non-governmental organisations (NGOs) took part in parallel fora (Adams, 2009). The central aim was to identify the principles of an agenda for action towards sustainable development in the future and the challenge was seen to require consensus at the highest level. A key outcome was the 'Agenda 21' document (extending to 40 chapters and 600 pages) detailing the issues, the actors and the means for achieving sustainable development by the start of the twenty-first century. Putting sustainable development into practice was seen to involve the participation of a full range of sectors, groups and organisations; in business and science, youth and church groups within communities and by local authorities as well as international agencies as seen in Figure 1.1. A number of important international conventions were also agreed at Rio, including the Convention on Biodiversity and the Framework Convention on Climate Change in recognition of the growing problems of sustainable use of ecosystems and of human-induced climate change. There was an optimism concerning a common interest on behalf of countries globally and between current and future generations that would drive sustainable development into practice.

Ten years later, 104 heads of state gathered again for the UN World Summit on Sustainable Development (WSSD) in Johannesburg, South Africa. The aim was to reinvigorate at the highest political level, the global commitment to a North–South partnership to achieve sustainable development. It has been referred to as 'by far the most inclusive summit to date' (Seyfang, 2003: 227) for the way

Figure 1.1 *The structure of Agenda 21*

Section 1: Social and Economic Dimensions

Eight chapters, covering international cooperation, combating poverty, consumption patterns, population, health, settlements and integrated environment and development decision-making.

Section 2: Conservation and Management of Resources for Development

Fourteen chapters on the environment. These cover the atmosphere, oceans, freshwaters and water resources, land-resource management, deforestation, desertification, mountain environments, sustainable agriculture and rural development. They also cover the conservation of biological diversity and biotechnology, toxic, hazardous, solid and radioactive wastes.

Section 3: Strengthening the Role of Major Groups

Ten chapters discussing the role of women, young people and indigenous people in sustainable development; the role of non-governmental organisations, local authorities, trade unions, business and scientists and farmers.

Section 4: Means of Implementation

Eight chapters, exploring how to pay for sustainable development, the need to transfer environmentally sound technology and science; the role of education, international capacity-building, international legal instruments and information flows.

Source: N. Robinson (1993).

in which more stakeholder groups were brought into formal meetings, including a bigger presence for business and many more NGOs from the developing world, representing issues of human rights, social justice and business accountability, for example. These activities suggested new ways of addressing sustainable development at a global level and a 'more decentralized understanding of where change comes from' (Bigg, 2004: 5). There was a new understanding of the complex interdependencies of environmental, social and economic development (Potter et al., 2008) and of the difficult political challenges of sustainable development. Key concerns at the start of the twenty-first century were for the continued degradation of environmental systems since Rio, but also for the persistence of poverty and evidence of widening global disparities.

In 2000, the UN community had committed to the achievement of eight Millennium Development Goals (MDGs) embracing many of these concerns. One of these goals refers explicitly to sustainable development and the actions of governments in preparing national sustainable development strategies, for example. Many others are central to sustainable development in that they demand better outcomes in the arenas that affect poorer groups. Figure 1.2 identifies the MDG goals, the specific targets set and the principal ways in which they link explicitly to the environment. However, at the WSSD, a central concern was the impacts of 'globalisation' on the poor. In short, whilst people and places were becoming more closely linked together within global markets and through flows of finance, for example, the benefits and costs of economic globalisation were not being shared equally across or within countries. Poverty, inequality and exclusion were identified as threats not just to the environment and economic prosperity but also to future security and democracy. Whereas globalisation had not been discussed at the Rio conference, a decade later it was central to understanding sustainable development as seen in the Johannesburg Declaration (Figure 1.3).

A further UN summit, 'Rio+20' is currently planned for 2012 to be held again in Brazil. All stakeholders have been invited not just to the conference, but to contribute in advance to a working document that will inform what is discussed and the official outcome documents. The aim of the conference is to secure renewed political commitment for sustainable development, to assess progress on the outcomes of previous summits and to address new and emerging challenges for the global community. Two themes are considered priorities: the challenges of moving to a 'green economy' and what the future institutional framework for sustainable development should be. What these challenges entail and how they have emerged is the focus of the next two chapters. In short, a number of crises are facing the world – of climate, economy, food and energy, and poverty for example – that are now understood as interlinked. The notion of the green economy seeks an economic system that can address and prevent these crises whilst also protecting the earth's ecosystems, provide economic growth and contribute to poverty alleviation. Questions of the future institutional framework for sustainable development include whether the formal organisations (including those within the UN) are 'fit for purpose' to guide, monitor and coordinate progress towards sustainable development in future. They also embrace many wider issues of 'governance' that confirm the important role across all scales, not only of governments, but also private business and civil society organisations in shaping the prospects for sustainable development. They also question the principles on which decisions are made, whether these are equitable and participatory, for example.

Evidently, sustainable development is considered a central and important challenge for international organisations such as the United Nations and for governments worldwide. It is seen to embrace linked

Figure 1.2 *The Millennium Development Goals and the environment*

Millennium Development Goal	Target	Selected environmental links*
1 To eradicate extreme poverty and hunger	1 To halve, between 1990 and 2015, the proportion of people whose income is less than US$1 a day 2 To halve, between 1990 and 2015, the proportion of people who suffer from hunger	Livelihood strategies and food security of the poor often depend directly on healthy ecosystems, and the diversity of goods and ecological services they provide. Natural capital accounts for 26 per cent of the wealth of low-income countries. Climate change affects agricultural productivity. Ground-level ozone damages crops.
2 To achieve universal primary education	3 To ensure that, by 2015, children everywhere, boys and girls alike, will be able to complete a full course of primary schooling	Cleaner air will decrease illnesses of children due to exposure to harmful air pollutants. As a result, they will miss fewer days of school. Water-related diseases such as diarrhoeal infections cost about 443 million school days each year and diminish learning potential.
3 To promote gender equality and empower women	4 To eliminate gender disparity in primary and secondary education, preferably by 2005 and in all levels of education no later than 2015	Indoor and outdoor air pollution is responsible for more than 2 million premature deaths annually. Poor women are particularly vulnerable to respiratory infections as they have high levels of exposure to indoor air pollution. Women and girls bear the brunt of collecting water and fuelwood, tasks made harder by environmental degradation such as water contamination and deforestation.
4 To reduce child mortality	5 To reduce by two-thirds, between 1990 and 2015, the under-five mortality rate	Acute respiratory infections are the leading cause of death in children. Pneumonia kills more children under the age of five than any other illness. Environmental factors such as indoor air pollution may increase children's susceptibility to pneumonia.

Goal	Target	
5 To improve maternal health	6 To reduce by three-quarters, between 1990 and 2015, the maternal mortality ratio	Water-related diseases, such as diarrhoea and cholera, kill an estimated 3 million people/year in developing countries, the majority of whom are children under the age of five. Diarrhoea has become the second biggest killer of children, with 1.8 million children dying every year (almost 5,000/day). Indoor air pollution and carrying heavy loads of water and fuelwood adversely affect women's health, and can make women less fit for childbirth and at greater risk of complications during pregnancy. Provision of clean water reduces the incidence of diseases that undermine maternal health and contribute to maternal mortality.
6 To combat HIV/Aids, malaria and other diseases	7 To have halted by 2015 and begun to reverse the spread of HIV/Aids 8 To have halted by 2015 and begun to reverse the incidence of malaria and other major diseases	Up to 20 per cent of the total burden of disease in developing countries may be associated with environmental risk factors. Preventative environmental health measures are as important as and at times more cost-effective than health treatment. New biodiversity-derived medicines hold promises for fighting major diseases.
7 To ensure environmental sustainability	9 To integrate the principles of sustainable development into country policies and programmes and reverse the loss of environmental resources 10 To halve, by 2015, the proportion of people without sustainable access to safe drinking water and basic sanitation 11 To have achieved, by 2020, a significant improvement in the lives of at least 100 million slum dwellers	Current trends in environmental degradation must be reversed in order to sustain the health and productivity of the world's ecosystems.

continued . . .

Figure 1.2 . . . continued

Millennium Development Goal	Target	Selected environmental links*
8 To develop a global partnership for development	12 To develop further an open, rule-based, predictable, non-discriminatory trading and financial system	Poor countries and regions are forced to exploit their natural resources to generate revenue and make huge debt repayments.
	13 To address the special needs of the least developed countries	Unfair globalisation practices export their harmful side-effects to countries that often do not have effective governance regimes.
	14 To address the special needs of landlocked countries and small island developing states	
	15 To deal comprehensively with the debt problems of developing countries through national and international measures in order to make debt sustainable in the long run	
	16 In cooperation with developing countries, to develop and implement strategies for decent and productive work for youth	
	17 In cooperation with pharmaceutical companies, to provide access to affordable, essential drugs in developing countries	
	18 In cooperation with the private sector, to make available the benefits of new technologies, especially information and communication	

Source: compiled from www.un.org/millenniumgoals/ and *UNEP (2007).

Figure 1.3 *The Johannesburg Declaration on Sustainable Development: the challenges we face*

- We recognise that poverty eradication, changing consumption and production patterns, and protecting and managing the natural resource base for economic and social development are overarching objectives of, and essential requirements for, sustainable development.
- The deep fault line that divides human society between the rich and the poor and the ever-increasing gap between the developed and developing worlds poses a major threat to global prosperity, security and stability.
- The global environment continues to suffer. Loss of biodiversity continues, fish stocks continue to be depleted, desertification claims more and more fertile land, the adverse effects of climate change are already evident, natural disasters are more frequent and more devastating and developing countries more vulnerable, and air, water and marine pollution continue to rob millions of a decent life.
- Globalisation has added a new dimension to these challenges. The rapid integration of markets, mobility of capital and significant increases in investment flows around the world have opened new challenges and opportunities for the pursuit of sustainable development. But the benefits and costs of globalisation are unevenly distributed, with developing countries facing special difficulties in meeting this challenge.
- We risk the entrenchment of these global disparities and unless we act in a manner that fundamentally changes their lives, the poor of the world may lose confidence in their representatives and the democratic systems to which we remain committed, seeing their representatives as nothing more than sounding brass or tinkling cymbals.

concerns for environmental degradation, poverty and exclusion currently and regarding the long-term viability of existing approaches in both environment and development. Figure 1.4 identifies the principles of sustainable development identified by the UK government. However, sustainable development is a term that has currency well beyond international organisations and heads of state, even becoming 'hard to avoid' (Gibson, 2005). It has been taken up extensively by a range of organisations and interests and is a term widely used in the media and across academic disciplines, for example. It is used to sell products, justify policy decisions and inspire action. For some, the term 'sustainable development' has been redefined so many times and used to cover so many aspects of society–environment relationships that there are 'doubts on whether anything good can ever be agreed' (Mawhinney, 2002: 1). For others, sustainable development is an idea that 'makes a difference' precisely because it requires debate and resolution of different interests and because it challenges both researchers and policy-makers alike (McNeill, 2000).

This chapter considers how and why sustainable development has become such a widespread but also contested notion. It identifies a number of different 'meanings' within different disciplinary

Figure 1.4 *The UK Government's Principles of Sustainable Development*

I.   Living within Environmental Limits
II.   Ensuring a Strong, Healthy and Just Society
III.   Achieving a Sustainable Economy
IV.   Promoting Good Governance
V.   Using Sound Science Responsibly

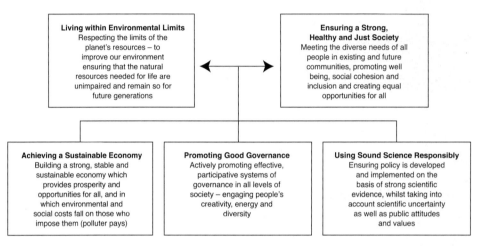

Source: Compiled from DEFRA (2005)

perspectives and considers changing ideas about how development can best be achieved and the actions required to manage environmental change. Identifying these changes in thinking and practice from which the notion of sustainable development has emerged is central to understanding the nature of current challenges and actions for sustainable development (the focus of Chapters 2 and 3 respectively).

The concept of sustainable development

Literally, sustainable development refers to maintaining development over time. However, there are possibly hundreds of definitions of the term currently in circulation, many divergent interpretations and thousands of variations applied in practice (Gibson, 2005). Figure 1.5 lists just a small number of such definitions and the varied interpretations of the concept. Definitions are important, as they are the basis on which the means (strategies, policies and mechanisms) for achieving sustainable development are built: how the human and environmental 'condition' is thought about, viewed or understood underpins subsequent interventions. As will be seen in this section,

Plate 1.1 *Promoting the messages of sustainable development*

a. Sign on entry to Kang, Botswana
Source: David Nash, University of Brighton.

b. Fresher's Fair, University of Brighton, England
Source: Elona Hoover, University of Brighton.

different disciplines have influenced and contributed to the sustainability debate, 'each making different assumptions about the relation between environment and the human subject' (Lee et al., 2000: 9) and assigning to it quite divergent orders of priority and recommendations in terms of policies, programmes and projects. Throughout this text, it will be apparent that although there are many shared concerns for, emerging principles of and evidence of progress towards sustainable development, there is also much uncertainty and contestation regarding how to best promote sustainable change and concerning the impacts of policies and mechanisms taken towards sustainable development.

Figure 1.5 *Defining and interpreting the contested concept of sustainable development*

Definitions of sustainable development

'In principle, such an optimal (sustainable growth) policy would seek to maintain an "acceptable" rate of growth in per-capita real incomes without depleting the national capital asset stock or the natural environmental asset stock.'

(Turner, 1988: 12)

'The net productivity of biomass (positive mass balance per unit area per unit time) maintained over decades to centuries.'

(Conway, 1987: 96)

'Development that meets the needs of the present without compromising the ability of future generations to meet their own needs.'

(WCED, 1987: 43)

'A sustainable society is one in which peoples' ability to do what they have good reason to value is continually enhanced.'

(Sen, 1999)

Interpretations of sustainable development

'Like motherhood, and God, it is difficult not to approve of it. At the same time, the idea of sustainable development is fraught with contradictions.'

(Redclift, 1997: 438)

'It is indistinguishable from the total development of society.'

(Barbier, 1987: 103)

'Its very ambiguity enables it to transcend the tensions inherent in its meaning.'

(O'Riordan, 1995: 21)

'Sustainable development appears to be an over-used, misunderstood phrase.'

(Mawhinney, 2002: 5)

As suggested in the quotations in Figure 1.5, the attractiveness of the concept of sustainable development may lie precisely in the varied ways in which it can be interpreted, enabling diverse and possibly incompatible interests to 'sign up to' sustainable development and to support a wide range of practical initiatives and causes. This is what is termed 'constructive ambiguity' in understanding the concept that enables a 'strategic flexibility' in terms of responses. However, it also confirms the need for ongoing critical consideration of whose values and interests are encompassed in particular kinds of policy and practical intervention and who may suffer costs and losses, that is, it is more than an academic debate. As Jacobs (1991) identified, sustainable development is a 'contestable concept', that like 'democracy' or 'equality' has a basic meaning that almost everyone is in favour of, but there are deep conflicts around how they should be understood and fostered.

Sustainable development as 'common sense'?

The challenges of understanding what the idea of sustainable development may mean, and what it entails in practice is evident in an analysis of the apparently simple definition provided by the WCED in 1987 (identified above in Figure 1.5). Common sense could suggest that development today should not be at the expense of future generations, but what is it that one generation is passing to another? Is it solely natural capital or does it include assets associated with human ingenuity and the application of technology, and what of language or other aspects of culture? The WCED definition also embraces difficult notions of limits and needs. What and how are limits set – by biophysical processes, technology or society, for example? What do we know of the 'needs' of future generations (or even the size and location of populations to come)? Quite evidently, 'needs' can mean different things to different people and change over time, linked to 'development' itself and society's ability to satisfy them. In one place, needs may be for open space and clean air, and in another, for material wealth even at the cost of greater pollution. How do we reconcile that needs in one place or amongst particular groups are often fulfilled at the expense of others? Or that development to date has enabled new needs to be defined within certain groups (that could be interpreted as 'wants') without satisfying the basic needs (increasingly understood as human rights) of others? These questions highlight some of the many sources of contestation in the debates over the meaning of sustainable development: conflict between the interests of present generations and those of the future; between human well-being and the protection of nature; between poor and rich; and between local and global. These questions frame the global challenges of sustainable development of Chapter 2.

Framing the concept of sustainable development

A number of frameworks or typologies have been proposed as a way of simplifying the evidently complex notion of sustainable development. Commonly, sustainable development is presented as three pillars, as seen in Figure 1.6. Such 'architectural metaphors' confirm the need to consider the social, ecological and economic arenas together and equally ('holistically') if the building is to

Figure 1.6 *Depictions of sustainable development*

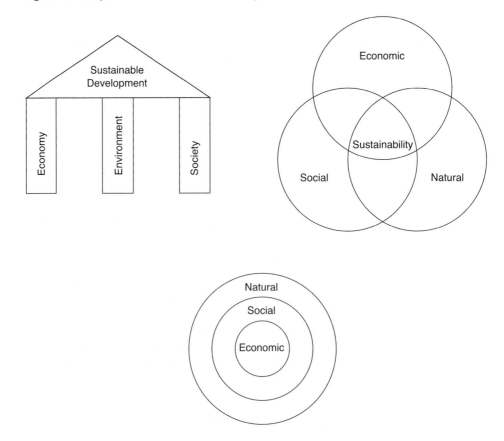

remain upright and development is sustainable. However, such depictions are less effective for communicating the interconnections between the 'uprights' and the need to integrate thinking and action in sustainable development across traditional disciplinary boundaries and established policy-making departments, for example. These principles of sustainable development are perhaps better portrayed by interlocking circles with sustainable development as where the circles intersect. Such depictions give attention to the objective of sustainable development as seeking to maximise the goals across all three spheres at the same time and the possibility of mutually supportive ('win-win-win') gains that can be made through sustainable development actions. The small area of overlap relative to the whole sphere portrays the unsustainable nature of much activity, but also opens the idea of the potential to expand this area of positive overlap. Importantly, this model supports understanding that achieving sustainable development in practice regularly involves trade-offs across the different spheres; that difficult choices have to be made at particular points in time and at particular scales as to what is being pursued and how; that certain goals can be compromised in the achievement of others; and that any action will carry unequal impacts for particular interests and for groups of people. This is the requirement for 'systems thinking' in sustainable development.

A further depiction of sustainable development is of concentric or 'nested' circles where the spheres of economy and society are shown as embedded in a wider circle of ecology. This portrays an understanding of environmental limits setting the boundaries within which a sustainable society and economy must be sought. This model presents a better illustration of how all human activities depend fundamentally on nature and portrays more clearly how activities that damage the functioning of natural systems ultimately weaken the basis of human existence itself.

Most recently, there have been suggestions that a further pillar sphere of sustainable development is required, that of cultural diversity as the root of a more moral, spiritual, ethical and sustainable way of life. This is argued by indigenous peoples in particular and within initiatives such as the Earth Charter that brings together a wide range of civil society organisations with shared concerns for the values underpinning more sustainable development: respect and care for cultural diversity as well as ecological integrity and for universal human rights and a culture of peace.

Disciplinary development of the concept

Whilst potentially all disciplines engage in some way with the concept of sustainable development, early contributions came largely from the 'environmental' spheres. As will be seen through this chapter, particular environmental disciplines (notably ecology and conservation biology) were especially prominent in important early work towards understanding emerging environmental problems and in designing measures to protect valued environments. These disciplines provided the evidence to support environmental movements and were instrumental in international legislation such as the Convention on Biological Diversity agreed at the Rio Earth Summit. Research from across the environmental sciences continues to inform major global studies such as the UN Millennium Ecosystem Assessment (2005) and in understanding climate change (discussed in Chapter 2). Developments in information communication technology have been very important in enabling the creation of vast databases and sophisticated modelling that are key to Earth Systems Science. This relatively new field of study is centred on integrating the physical, chemical and biological science disciplines to address the complexity and uncertainties of global environmental change and for understanding human activities as an integrated component and planetary-scale force in the Earth system (Steffen et al., 2004).

The disciplines of economics have also been very important in shaping the concept and practices of sustainable development. Their role in developing the notion of 'critical natural capital' is explored in Box 1.1. Ecological economists have also developed a range of means for applying economic calculations to environmental resources towards costing resources and resource functions both as inputs to economic activity and in terms of degradation and pollution. This work has underpinned a host of what are termed 'market-based mechanisms' towards achieving sustainable development in practice. These are considered more fully in a number of subsequent sections and chapters (as they are now widely used but are also heavily debated). Essentially, market-based mechanisms encompass measures to alter the economic costs of particular behaviours and production practices towards more sustainable outcomes. Simple examples are environmental taxes on petrol use and solid waste disposal that make these practices more costly to individuals and businesses.

Most recently, it is the third sphere of the models in Figure 1.6 that has accommodated much work in sustainable development.

Box 1.1

Forms of capital for sustainable development

For economists, various forms of 'capital' provide the capability to generate human well-being. These are human capital (skills, knowledge and technology, for example), physical capital (encompassed in housing, roads and machinery etc.) and natural capital (stocks of natural resources such as oil and minerals, but also biological diversity, atmospheric and hydrological cycles and such like). Economists query what forms of capital the concept of sustainable development requires us to consider for future generations; is it the 'overall stock' of human, physical and natural capital (suggesting that all forms of capital are equal and that a loss of wetlands could be compensated for by an increase in roads, for example)? Or are particular forms of capital more 'valuable', more important for well-being or more fundamental to future human survival?

Economists recognise that environmental resources are varied in terms of their location, volume and the functions that they provide for economic activity. Furthermore, environmental resources differ in terms of whether they can be 'substituted' or replaced by other resources and/or forms of capital. Hence, ecological economists differentiate between three categories of 'natural capital' in Figure 1.7. 'Critical natural capital' comprises those assets that cannot be recreated and are lost forever if degraded.

Figure 1.7 *Categories of natural capital*

Critical natural capital: capital that is required for survival. It can be viewed as functional (such as the presence of the ozone layer or the atmosphere in general) or valued (for example rare species valued in terms of their potential for health care).

Constant natural capital: capital that must be maintained in some form but can be adapted or replaced.

Tradable natural capital: natural capital which is not scarce or highly valued and which can be replaced.

The notion of critical natural capital (whilst still debated) has informed ideas of 'weak' and 'strong' sustainability in practice. A weak interpretation of sustainable development is where the total capital stock passed onto the next generation is constant or growing and all forms of natural capital can therefore be traded off and substituted with human capital. Strong sustainability demands the protection of critical natural capital because once lost, these assets are lost forever, and they cannot be recreated.

Source: compiled from Pearce et al. (1989); Barr (2008).

Discussions of sustainability as a political process have been taken up by a number of the social sciences centred on questions of power and outcomes for particular groups of people, across space and time. In short, this work raises sustainable development as a moral concept that seeks to define a 'fair and just' development (Starkey and Walford, 2001). The notion of 'environmental justice' is now a prominent part of contemporary discussions of the meaning and practice of sustainable development (see Walker, 2012). It points in particular to the distributional conflicts around the environment as outcomes of development that are occurring now and being disproportionately felt by some social groups within the current generation. The concern is for how environmental 'bads' (such as pollution and the degradation of ecosystem functioning) and 'goods' (including access to environmental resources that may be the material basis for livelihood or green space valued for health and recreational opportunities) are distributed across society. Environmental justice also encompasses a concern for the equity of environmental management interventions (who benefits and who loses through these) and for the nature of public involvement in decision-making.

Evidently, understanding the concept of sustainable development is itself a challenge. Gibson (2005: 39) suggests that 'out of the great diversity of theoretical formulations and applications, an essential commonality of shared concerns and principles can now be identified'. These are shown in Figure 1.8. It is clear that sustainable development embraces a rejection of things as they are in terms of the current patterns of environment and development globally. However, sustainable development is not an identifiable 'end point' or 'state' but requires ongoing critical consideration of the processes (the 'means') of development and decision making across all spheres of life. Hence, there is no blueprint for how to achieve sustainable development; rather the nature of sustainable development will be specific to particular places and points in time. However, the challenges of finding more positive alternatives to existing patterns and processes of development are universal and globally linked. In order to identify the challenges and opportunities of implementing sustainable development in practice, the following sections consider more fully the changes in thinking and practice from which the concept has developed. As Adams (2009: 26) suggests, sustainable development cannot be understood in 'an historical vacuum'.

Of particular importance are the changes in thinking about what constitutes 'development' and how best to achieve it, and changing

Figure 1.8 *The shared essentials of the concept of sustainability*

The concept of sustainability is:
- a challenge to conventional thinking and practice
- about long-term and short-term well-being
- comprehensive, covering all the core issues of decision-making
- recognition of the links and interdependencies, especially between humans and the biophysical foundations for life
- embedded in a world of complexity and surprise, in which precautionary approaches are necessary
- recognition of both inviolable limits and endless opportunities for creative innovation
- about an open-ended process, not a state
- about intertwined means and ends – culture and governance as well as ecology, society and economy
- both universal and context dependent

Source: Gibson (2005).

ideas about the 'environment' and actions required to manage environmental change. The following sections consider how certain ideas gained strength within particular political and economic contexts of the time (and in relation to the development and environment outcomes on the ground) to shape policies, strategies and solutions proposed. It will be seen that whilst the development and environment literatures in the past were substantially separate, the debates on sustainable development have been important in reshaping understanding and action in both these arenas and have brought them closer together.

Changing perceptions of development

> Poverty, hunger, disease and debt have been familiar words within the lexicon of development ever since formal development planning began, following the Second World War. In the past decade they have been joined by sustainability.
>
> (Adams, 2001: 1)

Development is often discussed in relation to 'developing countries', but is a concept which relates to all parts of the world at every level, from the individual to global transformations (Potter et al., 2008) and ideas about the best means to achieve these transformations are potentially as old as human civilisation. The origins of the modern era of international development as a planned activity and of development studies as a subject is suggested to link back to a speech made by President Truman of the US in 1949 when he employed the term 'underdeveloped areas' and identified poverty as a threat to prosperity and peace both within those areas and for the world as a whole. It was a time of post-war recovery in Europe for which the US administration had provided financial assistance through the Marshall Plan. Forty-four countries, largely from the industrialised world, had also come together at the Bretton Woods conference in 1944 to form the World Bank, the International Monetary Fund and the General Agreement on Tariffs and Trade. These new international organisations were designed to prevent the economic crisis and conflict that had characterised the previous period and ensure future economic stability, prosperity and a more peaceful world (see Willis, 2011 in this series).

Since these beginnings, the interdisciplinary field of development studies has seen many changes in thinking regarding the meaning and purpose of development (ideologies) and in development practice in the field (strategies of development). The following sections consider what in retrospect can be identified as key turning points in thinking about development that reshaped strategies in practice. Although these shifts are considered chronologically, rarely are theories totally replaced; rather, new ones find relative favour and contestation over the prescriptions for development flowing from them continue (Potter et al., 2008).

Optimistic early decades

During the 1950s and 1960s, development thinking (encompassing aspects of ideology and strategy) prioritised economic growth and the application of modern scientific and technical knowledge as the route to prosperity for countries considered 'underdeveloped' at that time. In short, the 'global development problem' was conceived as one in which less developed nations needed to 'catch up' with the West and enter the modern age of capitalism and liberal democracy. Underdevelopment was seen as an initial stage through which western nations had progressed and the gaps in development that existed could be gradually overcome through an 'imitative process' (Hettne, 2002: 7), through a sharing of the experience of the West in terms of capital and know-how. The process of development was seen in terms of modernisation and, in turn, modernisation was equated with westernisation (and an associated faith in the rationality of science and technology) during this period.

Insights from neo-classical economics as modelled by authors such as Rostow (1960) were very influential. Rostow suggested a linear path to economic development through a set of stages seen in Figure 1.9. His theory was based on the experience and history of the more developed societies (i.e. a Eurocentric stance) and his work offered the prospect of almost formulaic growth towards a modern economy and society that all countries could achieve if they followed a set of rules. Rostow's work was published at the height of the Cold War between the US and Russia and he himself was fiercely anti-communist. Industrialisation in a capitalist context was the central requirement in his model for moving from a traditional society based on agriculture to an urban-based society of high mass consumption. There was a key role for governments in stimulating local investment, demand and savings, in setting appropriate levels of taxes and ensuring that markets could operate efficiently. International assistance could also help countries reach the critical 'take-off' stage, after which the benefits of development and characteristics of modernisation would inevitably and spontaneously flow from the core to less-developed regions (both internationally and internally).

Figure 1.9 *The stages of economic development as modelled by Rostow*

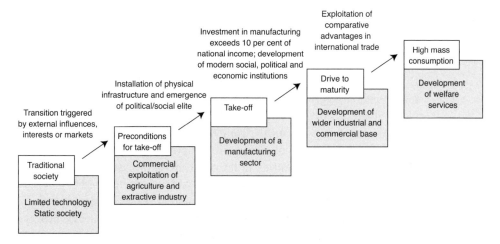

Source: adapted from Rostow (1960).

This 'modernisation thesis' dominated mainstream theories of economic development from the late 1950s through to the early 1970s. It was an optimistic time: it was thought that underdevelopment was due to constraints that were internal to these countries and could be solved quickly through the transfer of finance, technology and experience from the developed countries and through the spatial diffusion of modernity from the West and from urban centres to rural areas. There was the understanding of a linear, unconstrained path to economic development and an 'unswerving faith in the efficacy of urban-based industrial growth' (Potter et al., 2008: 94). Development planning in practice was characterised by large infrastructure, industrial and agricultural modernisation projects, mobilising new technologies and was increasingly financed through international borrowing. There was little concern for the environmental impacts of these policies on behalf of governments keen to deliver economic growth or amongst international donors and banks. The concern was 'grow now, clean up later' (Willis, 2011: 168).

The optimism of development theorists was generally not borne out by experience of development on the ground in that period. By the 1970s, inequality between and within countries had in fact worsened. The empirical evidence concerning economic growth as measured by gross national product (GNP) suggested that, whilst change had been achieved, 'development' was not shared equally amongst the populations of these nations. For example in Brazil in 1970, the poorest 40 per cent of the population received only 6.5 per cent of the total national income, in contrast to the 66.7 per cent of the total national income received by the richest 20 per cent of the population (Todaro, 1997). The optimism about a speedy end to underdevelopment started to fade.

During the late 1960s and 1970s, development thinking was influenced strongly by the writings of scholars within the developing world itself, particularly from Latin America and the Caribbean (notably those regions most strongly linked to the United States). They considered the socioeconomic structures and economic conditions of their countries in terms of the exploitative/dependent relations with other parts of the world, particularly through colonialism in the past and with the capitalist economy generally. The politics of development came to the fore within what became known as the radical or 'dependency' school of thought, most closely associated with the work of Andre Gunder Frank (1967), a European economist trained in America, but who carried out much research in Central and Latin America. In Europe too at this time, there was a reinvigorated interest in the work of Marx and an emerging 'New Left' movement that linked with the struggles of the Third World anti-colonial movements (Potter et al., 2008).

Fundamentally, the assertion in dependency theory was that underdevelopment was not the result of any inadequacies in economic, social or environmental conditions within those countries themselves, but the direct outcome of development elsewhere and the manner in which those countries were incorporated into the operations of the international capitalist system, that is, the structural disadvantages of these countries and regions. Rather than seeing the USA and Europe as the source of a cure for the ills of the developing world, dependency theorists saw the role of these regions as the origin of those ills, in actively creating the problems of underdevelopment. To use Frank's terminology, development and underdevelopment were two sides of the same coin. As illustrated in Figure 1.10, peripheral or satellite regions and countries are integrated into the world system through processes of unequal exchange and dependent relations with the metropolitan core. In consequence, the further entrenched they become in such processes, the more they are held back in development, rather than enabled to progress. This 'development of underdevelopment' was modelled as applying to processes of unequal exchange operating both internationally and internally within countries, and was used to explain patterns of regional and national underdevelopment in countries like Brazil.

The barriers to development as modelled by dependency theorists lay in the international division of labour and the terms of trade, rather than a lack of capital or entrepreneurial skills as within modernisation thinking. One of the principal policy responses to flow from these ideas was import substitution industrialisation (ISI). ISI is a strategy to enable peripheral countries to industrialise through looking inward (setting up domestic industry and supplying markets previously served by imports). It depends on a strong role for the state in protecting new industries via import tariffs and quotas and controlled access to foreign exchange. Many Latin American countries such as Brazil and Argentina had established substantial industrial bases by the 1960s using this strategy towards providing consumer goods such as clothing, cars, food and drinks to sizeable home markets. However, ISI proved less successful in relation to the

Figure 1.10 *The Frank model of underdevelopment*

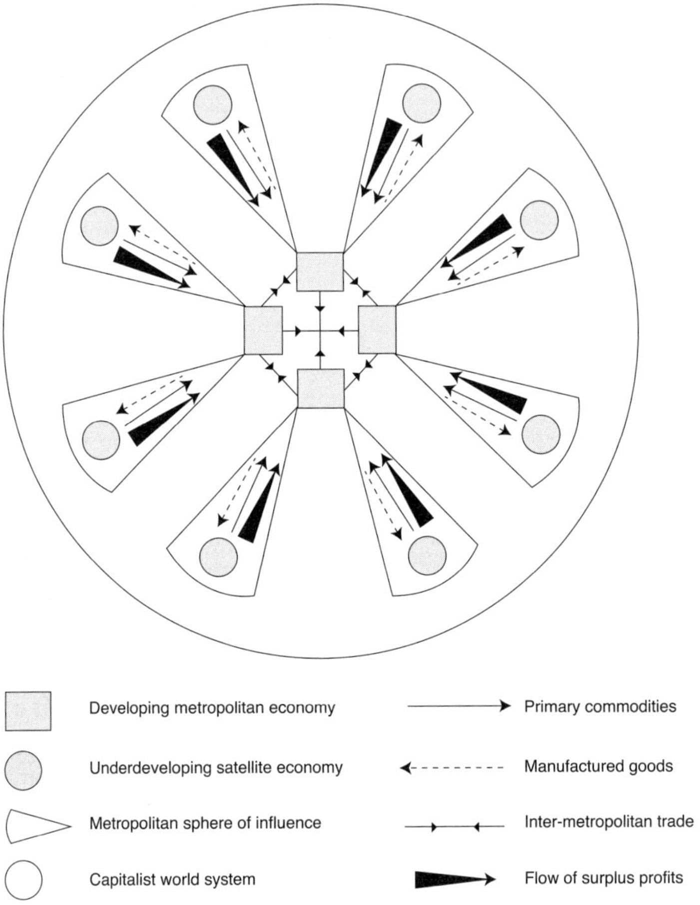

▢ Developing metropolitan economy	⟶	Primary commodities
◯ Underdeveloping satellite economy	⟵ - - - - -	Manufactured goods
▷ Metropolitan sphere of influence	→—←—	Inter-metropolitan trade
◯ Capitalist world system	▶	Flow of surplus profits

Source: adapted from Corbridge (1987).

production of intermediate and capital goods which were more capital- than labour-intensive (Hewitt, 2000). There were also internal problems including the lack of domestic capital to invest in such production and a lack of purchasing power on behalf of local, relatively poor, citizens. Other means towards 'withdrawal' from the international capitalist economy, such as through the formation of regional trading areas (as a means for expanding domestic markets), were not generally sustained over time. In short, dependency theory did much to expose the structural disadvantages of peripheral countries in relation to the capitalist core, and therefore how unlikely it was that they would follow the stages of economic growth mapped out on the basis of early experiences in Europe and North America (as modelled by modernisation theorists). However, the internal problems of local economies were generally underestimated within dependency theory. Critics also highlight that the fastest growing economies at that time, Japan and the 'Newly Industrialising Countries' of Hong Kong, Singapore, Taiwan and South Korea, were pursuing policies based generally on opening rather than closing their economies to external trade and export-led growth.

The lost decade of the 1980s?

By the 1980s, dependency theory gradually moved 'out of fashion', criticised in particular for its rather deterministic emphasis on the role of external economic structures in shaping society and development. Commentators were starting to consider the basic development conditions and needs of people within countries of the developing world, to focus on issues of self-reliance in development and on the internal forces of change. The expression 'another development' is often used as an umbrella term to include a broad sweep of changes in thinking regarding development and how best to achieve it from the late 1970s. Phrases such as 'growth with equity' or 'redistribution with growth' recognised that economic growth remained a fundamental component within development, but that it was critical to ensure that the benefits did not fall solely to a minority of the population. 'Development' became seen as a multidimensional concept encapsulating widespread improvements in the social as well as the material well-being of all in society.

In practice, these ideas meant that there was no single model or means for achieving development, rather it required investment in all sectors, across agriculture and industry and in health and education, for example. Diverse strategies (with a strong role for rural-based strategies) of 'development from below' (Stohr and Taylor, 1981) were envisaged, in contrast to single, 'top-down' models and urban-led growth in particular. It was asserted that development needed to be closely related to the specific local, historical, socio-cultural and institutional conditions of places, focused on mobilising the internal natural and human resources through appropriate technologies and giving priority to basic needs. In stark contrast to the theories of development up to that time, development in practice was to be more inclusive, with individual and cooperative actions and enterprises becoming the central means for (or 'agents' of) development. Strong notions of 'participatory development' emerged in recognition of the shortcomings of top-down, externally imposed and expert-oriented research and development practice (Cooke and Kothari, 2001). Furthermore, it started to be understood that development needed to be sustainable; it must encompass not only economic and social activities, but also those related to population, the use of natural resources and the resulting impacts on the environment.

The 1980s, however, have been referred to as the 'lost decade' in development. The suggestion is that with the exception of the 'Asian Tigers', the widespread experience in the developing world was of 'development reversals', that is, previous gains were lost and in many cases went into reverse. Table 1.1 confirms that world economic growth slowed through the decade, most markedly in developing regions and in particular in Africa and Latin America. In contrast, in India, China, Thailand, Malaysia, Philippines and Indonesia, for example, high levels of economic growth were reached. In combination with the earlier 'success stories' of the region, these experiences have been termed the 'East Asian miracle' (World Bank, 1993). The general 'openness' of these economies to foreign investment and a limited role of the state in economic activities has been widely used to explain the development performance of these countries relative to other developing regions.

For many developing countries through the 1980s, development had to be pursued in the context of global economic recession and a mounting 'debt crisis'. Starting in Latin America, with Brazil and Mexico announcing that they could no longer service their official

Table 1.1 *Economic growth rates in the world economy, 1971–2000 (annual growth rates in constant 1990 $US)*

GDP per capita	1971–80	1981–90	1991–2000
World	2.22	0.20	1.21
Industrialised countries	2.60	2.04	2.07
Developing countries	2.14	−0.17	1.04
Latin America	2.40	−0.83	1.26
Africa	1.04	−0.39	0.30
Asia	3.10	0.57	2.19
China	3.38	7.60	9.34
India	0.78	3.68	3.33

Source: compiled from Nayyar (2008) based on UN data.

debts, concern spread through the commercial banks and northern governments (that had previously lent huge monies in a context of low interest rates and global expansion) about widespread defaulting and the possible collapse of the international monetary system. Economic recession impacted on developing countries through a combination of declining international demand, increasing protectionism in the industrialised countries, deteriorating terms of trade, negative capital flows, continuing high interest rates and unfavourable lending conditions. These factors had serious implications for the environment, as considered in Chapter 3, and were primary aspects of the context in which sustainable development was pursued in the 1980s (and remain so). Not only did interest repayments mean money going out without any direct impacts on productive development internally, but savings had to be made, typically in the finance for environment departments and through cuts in social services. Domestic development policies were also increasingly shaped by the conditions laid down by the World Bank and the International Monetary Fund for further lending, as detailed below.

Through the 1980s there was also a considered 'impasse in development studies' (Schurmann, 2008). The suggestion was that 'old certainties' concerning understanding development were 'fading away' and that existing theories 'could ever less adequately explain experiences of development and underdevelopment' (p. 13), that is, there was a concern about how development was being theorised as well as the concerns over the increasingly diverse, inequitable and often negative experiences of development impacts on the ground. The end of the Cold War and the collapse of communism in the early 1990s undermined the strength of Marxist analyses that had underpinned dependency theories. A 'post-modern' critique within the social sciences generally at this time was also fundamentally about moving away from an era dominated by notions of modernisation and modernity. Furthermore, processes of 'globalisation' whereby countries and regions of the world were seen to be becoming more widely and deeply interconnected and ever more global in character, were also changing the position of the nation state and national governments across economic, social and political spheres. Yet the nation state was central to existing theories of development, as seen above. These factors raised many questions for those involved in both development thinking and practice through the 1990s.

The neo-liberal 1990s

From the late 1980s, disillusionment with the record of state involvement in the economy (and social life more broadly) mounted in the North. This was illustrated in the ascendancy of conservative governments and the politics of Reagan and Thatcher in the US and UK, for example. A belief in what Simon (2008: 86) terms the 'magic of the market' developed and neo-liberal ideas of development took hold. Neo-liberalism is essentially an approach to development that considers the free market to be the best way to initiate and sustain economic development. Typical policy implications of such an approach therefore centre on removing the influence of the state in markets; removing tariffs on imports and subsidies on exports, for example, and denationalising public industries and service provision. The roots of neo-liberalism are in the neo-classical economics of Adam Smith and 'this ideology rapidly became the economic orthodoxy in the North and was exported to the global South via aid policies and the measures formulated to address the debt crisis' (Simon, 2008: 87).

For many developing nations, their entry into the world economy through the 1990s was increasingly defined by the World Bank (WB) and the International Monetary Fund (IMF). As confirmed, many developing countries began to experience severe balance of payments difficulties in the 1980s that were considered to threaten the international financial system as a whole. Debt became the concern of the two 'mainstays of the global economic order', the WB and the IMF. The assessment was that the economic crisis in developing countries was more than a temporary liquidity issue (as it had been conceived in the 1970s). Rather, comprehensive, longer-term solutions were required, based on packages of broad policy reforms in indebted nations. The term 'structural adjustment programme' (SAP) is used to refer to the generic activities of the IMF and WB in this arena. The central objective of SAPs as defined by the WB was to 'modify the structure of an economy so that it can maintain both its growth rate and the viability of its balance of payments in the medium term' (Reed, 1996: 41), that is, to address issues of debt. Whilst each package of policy reform was tailored for the particular country, SAPs in practice included many or all of the elements listed in Figure 1.11. The significance of neo-liberal development ideas in shaping the policy prescriptions for economic restructuring and future development are clear; countries needed to increase the role of the market in their domestic economies and open their economies to

Figure 1.11 *The principal instruments of structural adjustment*

Cuts in:
- government expenditure
- public sector employment
- real wages

Pricing policies designed to:
- eliminate food subsidies
- raise agricultural prices
- cost recovery in public services

Trade liberalisation involving:
- currency devaluation
- credit reform
- privatisation of state-owned institutions
- higher interest rates

overseas investment and expand exports. The problems were considered to be internal to the indebted countries and the failings of past policies (particularly in terms of the over-involvement of the state in the economy).

It has been argued that the impacts of SAPs quickly went far beyond the original national contexts for which they were designed, to become an instrument for global economic restructuring (Reed, 1996) and through the conditions attached, they enabled the IMF and WB to 'virtually control the economies' of many developing nations (Hildyard, 1994: 26). The impacts of SAPs on economy, society and the environment have been widely documented and hotly debated (including from within the WB and IMF). Whilst they enabled countries through the period to access development finance (at a time when commercial lending and bilateral sources were restricted) and to achieve a degree of economic stability, they did not solve the challenges of debt. There is also a consensus that SAPs were 'blind' both to environmental and gender impacts and underestimated the social impacts particularly for most vulnerable groups. The impacts of SAPs on the prospects for sustainable development and the ways in which the WB has subsequently modified its policies are discussed in Chapter 3.

In short, what had been a very strong consensus within thinking on development started to unravel towards the new millennium.

Recognition grew of the unprecedented changes of a global character occurring in all arenas of economic, social, political (and environmental) activity and the term 'globalisation' became 'widely used to explain the causes and effects of most aspects of life at the turn of the century' (Willett, 2001: 1). Whilst global links and interconnections between places and peoples around the world had existed previously (through colonial ties or the spread of Islam, for example), globalisation was considered different from earlier periods including for the immediacy and intensity with which other parts of the world are now experienced within a 'speeded up' and 'hyper-connected' world. Current processes of global integration also now extend to much more than the flow of resources, people, goods and services, for example, as illustrated further in Chapters 2 and 3. However, globalisation was quickly seen to have an uneven reach. For example, Foreign Direct Investment (FDI) is a key driver of economic globalisation (Dicken, 2011). It refers to investments made overseas by companies (including setting up branches and investing in other firms). In 2000, such investments were largely occurring in other high income countries, as seen in Table 1.2, and the African continent could be suggested to be only loosely connected to this globalising economy. The rapid development and expansion of information communication technologies (particularly the internet and mobile phone technologies) has also been an important driver of economic (and cultural) globalisation. Yet, as illustrated in Figure 1.12, experience of this digital revolution has been uneven across world regions. Furthermore, understanding rose through the 1990s that not only were certain parts of the world being

Table 1.2 *Inward foreign direct investment, by major world region, 2000*

	Total flows, millions of dollars	% of world total
World	1,167,337	
Low income countries	6,812	0.6
Middle income countries	150,572	12.9
High income countries	1,009,929	86.5
East Asia and Pacific	42,847	4.0
Europe and Central Asia	28,495	2.4
Latin America and Caribbean	75,088	6.4
Middle East and North Africa	1,209	0.1
South Asia	3,093	0.3
Sub-Saharan Africa	6,676	0.6

Source: Compiled from World Bank (2003a).

Figure 1.12 *Uneven regional patterns of ICT: a) mobile phones per 1,000 people; b) internet users per 1,000 people*

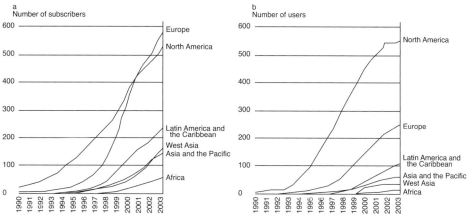

Source: UNEP (2007).

'left out', but processes of globalisation worked through existing patterns of uneven development to create increased differences between places (Castree, 2003).

One of the most radical reactions within development thinking to these dilemmas of development on the ground and to the limitations of neo-liberal ideas came from what is known as the 'post-development' school. Theorists such as Escobar (1995) consider development not in terms of structures of production or trade markets, but as a discourse and set of practices. Hence, they scrutinise the ideas, assumptions and power of particular representations of how societies and economies should develop. Within this thinking, the very idea of development as a desirable process itself is questioned given the ways in which it 'involves a dependent and subordinate process, creates and widens spatial inequalities, harms local cultures and values, perpetuates poverty and poor working and living conditions, produces unsustainable environments, and infringes human rights and democracy' (Hodder, 2000: 17). The 'development project' is also considered a 'failed' project (Rapley, 2008: 177) and is condemned for creating and producing the opposite of what it promised (Corbridge, 1999). In short, a post-development era depends on breaking the 'holds of westernisation' be it as organised by the aid industry or by the activities of western private capital. 'Defending the local' (such as through ecological, women's and people's organisations) and resisting the forces of globalisation are core prescriptions for change. Post-developmentalists emphasise grassroots participation and the capacities of organisations at the local level as agents of change (the suggestion being that the state has failed for the way that it has facilitated the westernisation of the development project). The post-development school has been criticised, among other things, for focusing on the 'worst' experiences of the last decades (Rigg, 1997) and of throwing 'the baby out with the bathwater' (Rapley, 2008: 180). However, it has helped reaffirm the importance of understanding difference and diversity and multiple paths to and patterns of development that are key principles for moving to more sustainable development processes, as seen in later chapters.

Rethinking development in the twenty-first century

By the end of the twentieth century, neo-liberal development ideas and policy prescriptions were increasingly challenged by outcomes seen in practice. In late 1996, for example, the decision by the government of Thailand to devalue their currency caused concern amongst international investors as to the overall stability of the Thai economy. This led to the withdrawal of large amounts of foreign capital and the development of a financial crisis that spread rapidly to countries in the wider region. The outcome was falling levels of economic growth and rising poverty in subsequent years. However, as Willis (2011) identifies, just as the Asian 'miracle' had many dimensions, so the experience of 'crisis' was different. Taiwan and Japan experienced little impact for example. Whilst there are competing explanations of whether the crisis was due to these Asian countries following the neo-liberal model too closely or not closely enough, it was confirmation of the diverse and increasingly volatile nature of economic development at the turn of the century. Neo-liberal prescriptions for development and the negative consequences of globalisation processes more widely were also being resisted politically. A growing 'anti-globalisation' movement drew together a wide array of groups including environmentalists, unions, peasant farmers, workers and feminists, in respect to what were understood as their shared concerns. Most visible were demonstrations at major meetings of those institutions such as the World Trade Organisation (WTO) and the World Bank that were central in promoting neo-liberalism. In 1999, for example, around 600,000 protesters and 700 different groups took to the streets in Seattle to disrupt the proceedings of the WTO conference.

The recent global economic crisis is a further source of much rethinking of previous development models. What started in late 2008 in the US as a financial crisis (within commercial banks and investment firms) quickly extended into a broader economic crisis affecting most countries globally. For many commentators, the crisis has undermined the credibility of the neo-liberal approach irreversibly. As the UK and US governments, for example, intervened to 'bail out' private banks (considered 'too big to fail' if the international financial system was to be preserved) and developed varied measures to stimulate and support domestic economies (through public finances), the suggestion is that this constitutes a 'dramatic reversal of the neoliberal philosophy that had come to dominate the political economy of most countries over the previous 30 years' (Knox and Marston, 2009: 10). Furthermore as Glennie (2009) suggests, these policy options (of state intervention in commercial markets on behalf of the public good) had effectively been denied (by those same governments) to many aid recipient countries through the conditions of structural adjustment programmes. A former Chief Economist at the WB suggests that developing countries are the 'innocent victims' of the crisis: 'those countries that are the most closely integrated into the global economy, having followed the advice of the international economic institutions and opened themselves up the most, have been amongst the worst hit' (Stiglitz, 2009: 40).

In some senses, development studies finds itself currently in a context similar to that of its origins; a new era of global cooperation is widely considered necessary to ensure global economic recovery that requires resisting a return to national protectionism (Stiglitz, 2009). Cooperation is also essential for addressing the (very much related) global environmental problems, including of climate change (Sachs, 2008; Giddens, 2009). The challenges for sustainable development within this current context are the focus of Chapter 2.

Changing perceptions of the environment

The history of environmental concern is quite similar to that of development studies: although people have held and articulated varying attitudes towards nature stretching back many years, it is since the 1960s that a coherent philosophy and language surrounding the environment ('environmentalism' as defined by Pepper in 1984) can be identified. In continuity with 'development thinking', it is possible to identify significant differences and changes over time concerning ideas about the environment; such as regarding humanity's fundamental relationship with nature, the influence of particular environmental disciplines and in terms of the prescribed conservation requirements. Although the focus here is largely on 'mainstream' environmentalism – that is, the broad consensus that can be identified as forwarded within successive conferences and publications of international institutions, for example – the diversity within this mainstream and at particular points in time should not be underestimated. As Martinez-Alier (2002: 1) suggests, 'not all environmentalists think and act alike'; among other things they draw support from different areas of environmental science such as conservation biology, environmental economics and political ecology, as seen through the next section. Box 1.2 illustrates a well-cited way of thinking through the differences that exist within environmentalism.

Development as environmentally destructive

In the 1960s, environmentalism was largely a movement reflecting European and American white, middle-class concerns. The undesirable effects of industrial and economic development were beginning to be seen via a number of 'conspicuous pollution incidents' (Bartelmus, 1994: 5) and people were worried about the effects on their own lifestyles and health: 'after two centuries of industrialism and urbanisation, people now began to rediscover the idea that they were part of nature' (McCormick, 1995: 56). Environmentalists in the US and Europe campaigned on issues such as air pollution and whaling and often received substantial support from the media. In contrast to earlier nature protection or conservation movements within these regions, environmentalism became overtly activist and political. The combination of actual changes in the environment and people's perceptions generally at this time brought widespread public support for the environmental movement, particularly amongst the younger groups. As Biswas and Biswas (1985: 25) suggested, 'the environment and Vietnam became two of the major issues over which youth rebelled against the establishment'. For the new environmentalists, it was not solely their local outdoor environments which were perceived to be under threat, but human survival itself. A number of very influential 'global future studies' were published in the early 1970s which served to reinforce

Box 1.2

Modes of thought concerning humanity and nature

It is argued that society's desire to manipulate nature, concomitant with an acceptance that the Earth nurtures our own existence, is inherent in the human condition. 'Technocentric' and 'ecocentric' refer to the two extreme positions. In reality, the distinction between these different perspectives is often blurred. As O'Riordan (1981) suggests, rarely is the world so neatly divided into two camps; rather we all tend to favour certain elements of both modes, depending on such factors as our changing economic status and the institutional setting or issue at hand. The categories should not, therefore, be thought of as rigidly fixed or mutually exclusive.

	Technocentric	**Ecocentric**
Environmental philosophies:	Human-centred: humanity has a desire to manipulate nature and make the world a more certain place in which to live.	Earth-centred: the Earth nurtures humanity's existence and should be treated with respect and humility.
Green labels:	'Dry Green'	'Deep Green'
	Reformist in that the present economic system is accepted, but considered to require some gradual revision.	Radical in that quite rapid and fundamental changes in economy and society are desired.
	Belief in political status quo, but more responsible and accountable institutions. Self-regulation through 'enlightened conscience'.	Supports devolved, political structures with emphasis on self-reliant communities and pursuit of justice and redistribution across generations.
Environmental management strategies:	Reliance on scientific credibility, modelling and prediction.	Management strategies geared to retaining global stability based on ecological principles of diversity and homeostasis.
	Promotes the appropriate manipulation of markets to create cost-effective solutions to environmental improvements.	New and fundamentally different conservation solutions required which are flexible and adaptable.
	Sustainable development through rational use of resources, better planning and clean technologies, for example.	Alternative and appropriate technologies.

Sources: compiled from Pepper (1996), O'Riordan (1981) and O'Riordan (1995).

and spread the fears and influence of western environmentalists. For example, texts such as *The Population Bomb* (Ehrlich, 1968), *Blueprint for Survival* (Goldsmith et al., 1972) and *The Limits to Growth* (Meadows et al., 1972) modelled an ever-expanding population and mounting demands of society on a fundamentally finite resource base.

Development and conservation were framed as incompatible: in conceiving the 'environment' as the stocks of substances found in nature, by definition these resources were considered to be ultimately limited and pollution and environmental deterioration understood as the inevitable consequences of industrial development. Based on this thinking, urgent and profound actions were required including population control in the developing world and 'zero-growth' in the world economy. The global future predictions gave little attention to the social, technological or institutional factors which affect the relationship between people and resources. Further to such environmental determinism, these reports were ahistorical in the sense that, for example, they gave no attention to how or why the world is divided into rich and poor. They were also apolitical in considering the future of the Earth (rather than people) as the

Plate 1.2 *The inevitable consequences of development? Industrial air pollution*

Source: Gordon Walker, Lancaster University.

overriding and paramount concern, with no consideration of how the solutions advocated would favour some nations or groups over others.

Not surprisingly, this environmental movement found little support in the developing nations; many such countries (outside Latin America) had only just gained independence and were sceptical regarding the motives behind proposals which seemed to limit their development objectives and remove sovereign control over resources. For developing nations (and their citizens) their development concerns were 'too little industry' rather than too much and were in stark contrast to the position of the more industrialised countries which used and consumed the bulk of resources and contributed most to the resulting industrial pollution.

In 1972, environmental issues became part of the international political agenda when 113 countries attended the UN Conference on the Human Environment in Stockholm. The conference has been referred to as a key event in the emergence of the discources of sustainable development and the integration of human development and environmental concerns as a global challenge. However, Adams suggests that 'it was only partly, and belatedly, concerned with the environmental and developmental problems of the emerging Third World' (2009: 59). The primary impetus for the conference had been the developed world's concerns about the international and cross-border effects of industrialisation (Sweden, for example, being particularly concerned about acid rain). In the event, the dialogue between government representatives, and within parallel meetings of NGOs at the conference, soon moved to wider issues including the relationship between environment and development issues. The term 'pollution of poverty' was used to refer to the environmental concerns of the poor, such as lack of clean water or sanitation, which threatened life itself for many in the developing world. It also encompassed the emerging recognition that a lack of development could also cause environmental degradation. However, detailed discussions of the links between poverty and environmental degradation or indeed the causes of poverty were limited at Stockholm. Whilst it was made clear that environment and development *should* be integrated, it was left unclear as to *how* this should happen (Adams, 2009).

Environmental scientists played a key role at this time in explaining environmental problems and providing the basis for policy measures

Plate 1.3 *The pollution of poverty*

a) Hazardous housing on a Kolkata roadside
Source: author.

b) Washing in the Jakarta floods
Source: author.

to address them. The principal responses (discussed more fully in Chapter 3) were through regulation and what have been referred to as 'command and control' instruments. To address and manage the detrimental 'side-effects' of economic development, national governments created a host of laws and regulations to specify the standards of pollution control that production processes and products had to meet. In this way, particular types of activity (such as discharging agricultural and industrial effluents into water courses) could be controlled, harmful substances (such as DDT) were banned from use and industry (the largest polluter at the time) was made to invest in the technologies to reduce polluting emissions. Through the 1970s and 1980s, an increasing number of international agreements and treaties based on similar regulatory mechanisms were also developed (termed 'Multilateral Environmental Agreements', MEAs). These set the principles and rules for how individual states should behave in relation to cross-boundary environmental problems (such as acid rain but also including trade in certain materials and species), in relation to common property resources (such as oceans and fisheries) and in relation to public goods (essentially the atmosphere). One of the principal outcomes of the Stockholm conference was the formation of the UN Environment Programme (UNEP), a new intergovernmental body to focus environmental action internationally and to serve as an 'international environmental watchdog' and monitor global environmental change. UNEP continues to oversee the development of MEAs. However, there are problems in negotiating and enforcing this regulatory approach to environmental policy making, as discussed in Chapter 3.

Conservation as dependent on development

By the late 1970s, there was some evidence that the two previously separate arenas of environment and development were starting to be seen as interconnected concerns and as global responsibilities. The challenge for the 1980s was to translate the substantial rhetoric into policies and actions in practice. A key document was the World Conservation Strategy (WCS) in 1980, referred to as the 'launchpad' for the concept of sustainable development (Mather and Chapman, 1995: 248). The WCS was compiled by the International Union for the Conservation of Nature and Natural Resources (IUCN), historically a body concerned with wildlife conservation and nature protection. However, the WCS emphasised the many ways in which human survival, future needs in science, medicine and industry are linked to ecological processes and genetic diversity, and urged the sustainable use of species and ecosystems. In addition, development for the first time was suggested as a major means for achieving conservation rather than being detrimental to it. However, for many commentators, the WCS remained principally a 'conservation document' and shared the limitations of earlier global future studies, including its portrayal of a very neo-Malthusian future in which escalating numbers of people were identified as the source of short-sighted approaches to natural resource management. Little attention was paid to the political, social, cultural or economic dimensions of resource use and how these also shape the prospects for action. The failure of what was proposed as a global agenda for environmental action to recognise the political nature of the development process was particularly problematic. As Reid (1995: 43) states:

> To suggest solutions based on such 'self-evident' principles as 'population increase must be halted' and 'carrying capacity must be respected' without considering the political realities affecting the chances of their being implemented is to ignore the rights of local people, neglect the impact of social and economic forces on their access to and use of resources, and deny the reality of the effort needed to surmount the many political difficulties associated with negotiating 'improvements' that have recognisable similarity with 'obvious' global solutions.

As identified at the outset of this chapter, the UN established the World Commission on Environment and Development to identify the long-term environmental strategies for the international community. Prior to the publication of their report in 1987, the urgency of finding actions at a global scale was confirmed. Scientists in 1985, for example, quite unexpectedly discovered a hole in the ozone layer over Antarctica and in the following year, the breakdown of the nuclear reactors at Chernobyl in the Ukraine drew global political attention as the contamination spread far beyond national boundaries. The 1987 WCED report extended the ideas of sustainable development significantly beyond those of the WCS. The WCED started with people (rather than the environment) and gave greater attention to human development concerns and the kinds of environmental policy needed to achieve these: to the challenge of overcoming poverty and meeting basic needs and to integrating the environment into economic decision-making. It was underpinned by a strong ideology that nature and the Earth *could* be managed (i.e. there was no spectre of economic disaster as portrayed by the WCS or in earlier scenarios) and for this, reviving economic growth was considered essential. No longer was it a question of whether development was desirable: economic growth was central to the Commission's proposals for environmental protection and 'degraded and deteriorating environments were seen to be inimical to continued development' (Mather and Chapman, 1995: 248). But new forms of economic growth would be key to sustainable development: growth must be less material and energy intensive, generate less pollution and waste and be more equitably shared.

The WCED report was also central in putting questions of intragenerational equity into the sustainability discourse, that is, the immediate development and environment concerns particularly in developing countries and for poorer groups in society specifically. The report stressed the futility of addressing environmental problems without tackling the factors underlying world poverty and international inequality within the current generation. Equity concerns were also evident in the recognition of differentiated responsibilities for current environmental impacts and in relation to the changes in action that are required:

> an additional person in an industrial country consumes far more than an additional person in the Third World. Consumption patterns and processes are as important as numbers of consumers in the conservation of resources.
>
> (WCED, 1987: 95)

Similarly, in relation to the objective of a changed quality in economic growth in the future, the Commission acknowledged varied responsibilities: 'if industrial development is to be sustainable over the long term, it will have to change radically in terms of the quality of that development, particularly in industrialised countries' (WCED, 1987: 213).

There is little doubt that the WCED (and the Rio conference held to assess progress five years on) have been very important in shaping the ideas and practices of sustainable development. However, critics focus on the 'comfortable reformism' of the WCED report, that is, the continued prominence given to economic growth and the suggestion that it can be reconciled with environmental conservation without any significant adjustments to the prevailing capitalist system. In turn, the substantial political and economic changes required in future to achieve sustainable development were underestimated (Starke, 1990). Adams (2009: 94) suggests that the

Rio agenda remained dominated by northern priorities focused on issues of the global commons such as biodiversity and climate change (argued to be more amenable to political and technical solutions than poverty or global inequality) and was essentially about 'tuning the economic machine, not redesigning it'. Furthermore, the principal policy solutions within the WCED report continued to emphasise large-scale, government-led initiatives, particularly in agriculture and renewable energy, to achieve sustainable growth with little reference to the role of the private sector or 'market mechanisms' more widely in environmental policy making. This can now be considered 'out of step' with how economic growth was actually occurring through the 1990s, as seen in the sections above. There was certainly substantial optimism at the Rio conference that new global commitments between North and South could be achieved (in the political-economic context of the end of the Cold War and recovery from economic recession). However, many of the problems of environment and development identified at Rio got worse through the 1990s and the finances identified as necessary to implement Agenda 21 generally did not materialise.

Market environmentalism and ecological modernisation

Through the 1990s, environmental thinking and action became increasingly framed by economic discourses. This was in part due to the rising interest of the disciplines of economics in issues of environmental resource use, but was substantially a reflection of the mounting influence of neo-liberal thinking through this period, whereby the extension of market relationships into ever more arenas of society was considered the best way forward. 'Market environmentalism' is a broad set of ideas centred on the market as the most important mechanism for regulating human use of the environment. Key ideas include that as market signals change (such as when resources become more scarce, 'competition' increases and prices rise), use generally will be discouraged and there is the economic incentive to develop substitutes and raise efficiency in use. Putting a price on environmental goods and services is also key to ensuring that markets can then operate for these and 'rational' decisions on use and management can be made. Under this 'free-market' thinking, privately owned resources are argued to be more efficiently managed and conserved, whilst open-access resources are liable to over-exploitation and degradation.

'Ecological modernisation' emerged in the 1980s, particularly in the Netherlands and West Germany. It can be considered an extension or sub-set of market environmentalism (Adams, 2009). This thinking, for example, considers the market as central in delivering sustainable development, but concedes that substantially 'free' (or what is termed 'unfettered') market capitalism does have negative environmental impacts and carries ecological dangers. It embraces a belief that markets can, however, be made to work better for the environment, that capitalism can be shaped to reduce such negative outcomes and can be made to have positive environmental improvements. Central to the considered opportunities of ecological modernisation (seen in Figure 1.13) are innovations in improved technologies, products and production processes and the greening of business and corporations. Ecological modernisation also recognises the role of improved planning and environmental assessment procedures and regulations that address problems of pollution at their source. Evidently it embraces a 'technocentric philosophy' as identified in Box 1.2 in the pursuit of 'rational, technical solutions to environmental problems and more efficient institutions for environmental management and control' (Adams, 2009: 126). It is optimistic in that institutions can change, people within them learn and future industrial development will overcome the global ecological challenges. Furthermore, everyone participates and benefits, through cooperation between industry, scientists, governments and environmental groups.

This kind of thinking has profoundly influenced strategies and actions in sustainable development into the twenty-first century, as considered in Chapter 3. For some, this reflects the way that market environmentalism and ecological modernisation offer strategies for sustainable development based on making only minor adjustments to conventional growth models that in turn explains their popularity with politicians and business for example. Critics question the faith put in the power of consumers to influence business, manufacturing and retailers to operate in more environmentally benign or indeed positive ways. For example, whilst purchases of 'fairly traded' and organic produce are expanding (see Chapter 3), there are concerns whether such 'greener consumption' can be 'scaled up' to make the extensive changes required (and whether lower income consumers can afford it). There are also more fundamental questions of whether 'shopping can save the planet', that is, whether society can consume its way out of environmental crisis. For many, it is the relative

Figure 1.13 *The opportunities of ecological modernisation*

- The dematerialisation of production processes to make them more energy and resource efficient, i.e. reducing the energy/resources for each unit of output, will lead at an aggregate level, to a decoupling of economic growth and resource use.
- Improvements in income and standards of living thereby become less dependent on natural resource inputs and result in less environmental degradation.
- Pollution prevention pays: businesses reduce costs through technological changes that reduce waste and pollution.
- Large scale, 'smoke-stack' industries that cannot be made ecologically sound are gradually phased out.
- Markets for pollution abatement equipment and other 'green technologies' are growing.
- Green consumerism is stimulating demand for goods that minimise environmental damage in the way that they are produced and that have reduced environmental impact when used.
- Science and technology support the integration of environmental considerations into the design, production and final disposal of all products such as through integrated product policies and life cycle assessment.
- Businesses take account of longer-term, environmental outcomes such as through environmental management systems and are encouraged to do so through government taxes and permits that penalise environmentally damaging activities.
- Markets have a key role in the transmission of both ecological practices and ideas. Knowledgeable consumers encourage manufactures and retailers to advertise the environmental credentials of their products, processes and outcomes.

Source: compiled from Carter (2007).

silence of this thinking on important North–South issues such as international trade and globalisation and social justice concerns that are the principle limitations of ecological modernisation thinking.

Environmental Justice and environmentalism of the poor

In the early 1990s, a critique of environmentalism gathered pace (particularly in the US) that charged the 'mainstream' environmental movements and organisations for being both insufficiently activist, and racist and classist (Pezzullo and Sandler, 2007). The early 'Environmental Justice' (EJ) movement focused on urban-based injustices to people, pollution, hazardous waste and the environmental dangers that were seen to be spatially concentrated in poor and minority neighbourhoods (Walker, 2012). The movement challenged the major environmental organisations to expand their agendas into issues of community health and urban community development and to be more accountable to poorer communities both in the US and internationally. Importantly, the movement grew out of particular cases of local people themselves affected by these issues, taking action and bringing these to public attention (rather than being led by environmental professionals as in mainstream environmentalism). One case that received high profile media attention (including being the basis of the film *Erin Brockovich*) was at 'Love Canal', a suburb within Buffalo, New York State in the US. In the late 1970s, the health impacts of toxic chemicals buried by a plastics company two decades earlier started to be seen in widespread miscarriages and birth deformities within neighbouring low income and largely coloured communities. Lois Gibbs was a resident of the area concerned about the constant ill-health of her son and linked it to the location of his primary school. She collected information on illnesses within the locality, organised petitions and lobbied the authorities to undertake environmental testing. Further widespread public action and pressure on local politicians led to the eventual declaration of the area as unsafe for human habitation and to the relocation of families.

In 1991, more than a thousand activists from the US, Canada and Central America came together at the First National People of Color Environmental Leadership Summit. The 17 principles of Environmental Justice listed in Figure 1.14 that were adopted at the summit are widely considered to be the defining document for the EJ movement. As seen, the principles confirm a much expanded conception of environmental issues to include a range of health, spiritual and education concerns. The strongly activist stance and call for efforts to secure justice for all peoples (not only racial justice) is also seen.

Concurrently through the 1990s, environmental concern and action was being expressed through a growing range of movements within developing countries. Martinez-Alier (2002: 14) identifies 'environmentalism of the poor' as growing out of thousands of 'local, regional, national and global ecological distribution conflicts and caused by economic growth and social inequalities'. Their action and concern centres on the impacts of ecological conflict on the livelihoods of peasant farmers, pastoralists and fishing communities, for example, who are often poor and in many of these countries, comprise the majority of the population (Martinez-Alier, 2002). As with the EJ movement, environmentalism of the poor emphasises what are considered the inevitable distributional conflicts of continued economic growth and persistent inequality and the geographies of the environmental and social impacts being felt currently. It therefore challenges not only mainstream environmental thinking but also the prevailing development models for the way in

Figure 1.14 *The principles of Environmental Justice, First National People of Color Environmental Leadership Summit, 1991*

Environmental Justice:

- affirms the sacredness of Mother Earth, ecological unity and the interdependence of all species, and the right to be free from ecological destruction.
- demands that public policy be based on mutual respect and justice for all peoples, free from any form of discrimination or bias.
- mandates the right to ethical, balanced and responsible uses of land and renewable resources in the interest of a sustainable planet for humans and other living things.
- calls for universal protection from nuclear testing, extraction, production and disposal of toxic/hazardous wastes and poisons and nuclear testing that threaten the fundamental right to clean air, land, water and food.
- affirms the fundamental right to political, economic, cultural and environmental self-determination of all peoples.
- demands the cessation of the production of all toxins, hazardous wastes, and radioactive materials, and that all past and current producers be held strictly accountable to the people for detoxification and the containment at the point of production.
- demands the right to participate as equal partners at every level of decision-making, including needs assessment, planning, implementation, enforcement and evaluation.
- affirms the right of all workers to a safe and healthy environment without being forced to choose between an unsafe livelihood and unemployment. It also affirms the right of those who work at home to be free from environmental hazards.
- protects the right of victims of environmental injustice to receive full compensation and reparations for damages as well as quality health care.
- considers governmental acts of environmental injustice a violation of international law, the Universal Declaration on Human Rights, and the United Nations Convention on Genocide.
- must recognise a special legal and natural relationship of Native People to the US government through treaties, agreements, compacts and covenants affirming sovereignty and self-determination.
- affirms the need for urban and rural ecological policies to clean up and rebuild our cities and rural areas in balance with nature, honouring the cultural integrity of all our communities, and provide fair access for all to the full range of resources.
- calls for the strict enforcement of principles of informed consent, and a halt to the testing of experimental reproductive and medical procedures and vaccinations on people of color.
- opposes the destructive operations of multi-national corporations.
- opposes military occupation, repression and exploitation of lands, peoples and cultures and other life forms.
- calls for the education of present and future generations which emphasizes social and environmental issues, based on our experience and an appreciation of our diverse cultural perspectives.
- requires that we, as individuals, make personal and consumer choices to consume as little of Mother Earth's resources and to produce as little waste as possible; and make the conscious decision to challenge and reprioritise our lifestyles to ensure the health of the natural world for present and future generations.

Source: compiled from http://www.ejnet.org/ej/principles.html (accessed 24 February 2012).

which they encompass growing requirements for natural resources and consumption goods and involve the continued extension into new geographical frontiers (typically within poorer countries) such as through palm oil or GM crop technologies. The EJ movement also highlights the diverse environmental concerns of the poor that are often associated with immediate material livelihoods and human well-being and the importance of issues of indigenous rights, spiritual and cultural values and the sacredness of nature in defending those livelihoods (Martinez-Alier, 2002). Work in the discipline of political ecology has done much to inform understanding of the environmentalism of the poor, particularly through exposing the (unequal) power relations encompassed in accessing and managing environmental resources (see Chapter 2).

Substantial confirmation of the importance of these distributive and justice concerns for future sustainable development came in 2005 with the publication of the findings of the Millennium Ecosystem Assessment. The MEA aimed 'to assess the consequences of ecosystem change for human wellbeing and to establish the scientific basis for actions needed to enhance the conservation and sustainable use of ecosystems and their contribution to human wellbeing' (MEA, 2005: ii). It was a huge undertaking, drawing information from local communities and practitioners as well from the scientific literature. In its preparation, four working groups and over 2,000 reviewers from 95 countries were involved, guided by a Board that had representatives of five international conventions, five UN agencies, international scientific organisations, governments, private sector representatives, NGOs and indigenous groups. The significance of the MEA in understanding the contemporary challenges of sustainable development are considered further in Chapter 2. In its summary, the MEA identified that 'three major problems associated with our management of the world's ecosystems are already causing significant harm to some people, particularly the poor, and unless addressed will substantially diminish the long-term benefits we obtain from ecosystems' (MEA, 2005: 1). Two of these three findings in Figure 1.15 explicitly embrace the concerns of environmental justice and environmentalism of the poor. The assessment confirms that the degradation of ecosystem services is already a significant barrier to the achievement of the Millennium Development Goals and many of the regions facing the greatest challenges in achieving these goals coincide with those facing significant problems of ecosystem degradation.

Figure 1.15 *The key findings of the Millennium Ecosystem Assessment*

1. Approximately 60% of the ecosystem services examined are being degraded or used unsustainably; including fresh water, capture fisheries, air and water purification and the regulation of regional and local climate, natural hazards and pests. Many ecosystem services have been degraded as a consequence of actions taken to increase the supply of other services, such as food. *These trade-offs often shift the costs of degradation from one group of people to another or defer costs to future generations.*

2. There is evidence that changes in ecosystems are increasing the likelihood of nonlinear changes in ecosystems (including accelerating, abrupt and potentially irreversible changes) that have important consequences for human well-being. Examples include the creation of 'dead zones' in coastal waters and shifts in regional climate.

3. The harmful effects of the degradation of ecosystem services (the persistent decreases in the capacity of an ecosystem to deliver services) *are being borne disproportionately by the poor, are contributing to growing inequities and disparities across groups of people and are sometimes the principal factor causing poverty and social conflict.*

Source: compiled from UN-DESA (2005); *emphases added.*

Conclusion

Evidently, the idea of sustainable development is not new but has a substantial history. What was new in the 1980s was the way in which the two literatures of development and environmentalism started to come closer together, recognising the significant and interdependent nature of their goals and agendas. Into the 1990s, the concept of sustainable development encapsulated notions of development based in the reality of local environments and substantially expanded notions of the environment and the functions it plays in human societal development. Importantly, sustainable development was recognised as a *global* challenge: ultimately, the achievement of environment and development ends in any single location or for any group of people is connected in some way to what is happening elsewhere, for others. By the turn of the millennium, the world itself was characterised by unprecedented rates and degrees of economic, political, social and environmental change and the understanding of sustainable development came to encompass the challenges and opportunities presented by a globalising world. In the second decade of the new century, the concept of sustainable development now frames the debates concerning climate change, global economic recession and the persistence of poverty. Not only are these understood as interlinked challenges of sustainable development, but the concept of sustainable development now embraces a comprehensive critique of the 'governance' of environment and development that includes substantial rethinking of the processes of decision making and the values underpinning the actions of international institutions, governments, bankers and individual consumers. Further details of how these debates (and the rapidly changing context for sustainable development) are shaping the challenges of sustainable development and the nature of actions taken by the varied agencies in development are considered in the following two chapters. The outcomes of these in rural and urban areas are the focus of Chapters 4 and 5.

Summary

There are many different definitions of sustainable development coming from various disciplines and with different assumptions about the basic relationship between society and nature.

Ideas of sustainable development have a long history in the literatures of both development and environmentalism.

There have been a number of important international conferences within which actions towards sustainable development have been debated (and contested) at the highest levels of government.

Ideas concerning the best way of achieving development and managing environmental resources have changed over time, but are rarely replaced entirely. Neo-liberalism and ecological modernisation have dominated mainstream ideas in recent decades but are not subscribed to by all interests, and are increasingly challenged (in theory and practice).

Sustainable development is currently pursued in the context of a set of inter-linked global 'crises' of climate, economy and persistent poverty.

Discussion questions

- Select a country and consider the prospects for achieving the Millennium Development Goals by 2015 in that country.

- How has the role of governments in promoting more sustainable development changed in the last two decades?

- What are the principal ways in which debt challenges the prospects for sustainable development as defined by the WCED?

● How important are 'mega-summits' for progress towards more sustainable development?

Further reading

Adams, W.M. (2009) *Green Development*, third edition, Routledge, Abingdon. This is a more advanced text but provides a thorough review of both mainstream ideas in environment and development thinking and how these have been challenged, including from within the developing world.

Potter, R.B., Binns, J.A., Elliott, J.A. and Smith, D. (2008) *Geographies of Development*, third edition, Pearson Education Limited, Harlow. A good overview text for understanding the major issues in development.

WCED (World Commission on Environment and Development) (1987) *Our Common Future*, Oxford University Press, Oxford. The landmark text that first defined the concept of sustainable development and identified the linked challenges of environment and development for the global community. It continues to shape how sustainable development is understood and pursued in practice.

Willis, K. (2011) *Theories and Practices of Development*, second edition, Routledge, Abingdon. A very accessible and readable introductory text for understanding the key development theories and how these have shaped development interventions in practice.

Websites

www.unstats.org   An important source for many internationally compiled and authoritative statistics including regarding progress on the Millennium Development Goals.

www.maweb.org   Home page for the Millennium Ecosystem Assessment and the extensive research and reports regarding the current condition and trends of the world's ecosystems and the services they provide and the consequences for human well-being.

www.worldbank.org   Host to extensive data, resources, reports and publications concerning development including country-level Poverty Reduction Strategy Papers and the annual *World Development Reports*.

www.uncsd2012.org   Official website for the forthcoming (2012) UN Conference on Sustainable Development, 'Rio +20'.

② The global challenges of sustainable development

Learning outcomes

At the end of this chapter you should be able to:

- Identify the key global challenges of sustainable development
- Appreciate that the nature of the challenges and opportunities for sustainable development are socially defined and locally distinct
- Realise how questions of power, poverty and inequality are central to understanding notions of resource 'scarcity' and environmental 'limits' to development
- Appreciate the many dimensions of poverty and the complex linkages between poverty and environmental resources the functions and services provided for human welfare
- Identify the key features of the challenges presented by climate change for sustainable development

Key concepts

Environmental limits; resource scarcity; low-carbon development; resource-curse; poverty; ecosystem services; political ecology; climate warming.

Introduction

Chapter 1 showed that ideas about how best to achieve development and about society's relationship with the environment changed significantly through the last decades of the twentieth century, particularly as evidence emerged 'on the ground' about the successes and failures therein. Most recently, the global challenges of finding patterns and processes of development that are more sustainable have been exposed starkly by the global economic crisis. Previous trends of globalisation and continuing economic growth that 'many were starting to see as unstoppable' (Forum for the Future, 2010: 17) have been disturbed. There is much debate as to whether this new context can provide unique opportunities for sustainable development and for the 'world to reconstruct itself' (UNICEF, 2009) or whether it will threaten progress made.

This chapter outlines some of the key global challenges of sustainable development. Subsequent chapters identify how these are being taken on in practice, across spatial scales and sectors. Many of the core dilemmas in sustainable development are seen to be long standing including the resource dependent nature of development, the persistence of poverty and rising inequality in terms of access to development opportunities and environmental improvements. The chapter highlights where new research is shaping understanding of these 'old' challenges. For example, successive reports of the Intergovernmental Panel on Climate Change (IPCC) have now confirmed the anthropogenic causes of climate change (see Figure 2.1). Work in both poverty analysis and ecosystem change has also provided more holistic understandings of the linkages between environments and human welfare that is now central to strategies for meeting the MDGs, for example. New challenges for sustainable development are identified, among them those presented by the rapid development in the last decade of populous countries including China and India. Whilst this has delivered substantial benefits (removing over 400,000 people from extreme poverty, for example), higher standards of living and increased industrial production bring raised resource demands and enhanced carbon emissions with global impacts. The extent and pace of these economic transformations are unprecedented, as is the economic crisis. Both bring the global challenge of moving to lower carbon futures into sharp relief.

The emphasis in this chapter is on the challenges of sustainable development for the global community. However, the nature of these challenges and the opportunities for sustainable development will ultimately be manifested spatially and locally defined. Climate change impacts, for example, are hugely varied by physical location,

Figure 2.1 *Human impact on climate warming confirmed by the Intergovernmental Panel on Climate Change (emphases added)*

1995: The balance of evidence suggests a *discernible human influence* on global climate

2001: *most of the warming* observed over the last 50 years *is attributable* to human activities

2007: warming of the climate is *unequivocal*. Most of the observed increase in globally averaged temperatures since the mid twentieth century is *very likely due* to the observed increase in *anthropogenic* greenhouse gas concentrations.

the capacity to undertake adaptation measures, the structure of a particular economy and even within households. Similarly, global projections for population, for future resource consumption and for economic recovery all depend on a wide range of decisions and actions at various scales (the focus of Chapter 3) that will shape the integrated economic, environmental and social outcomes of development in practice (Chapters 4 and 5). Fundamentally, people live in places and it is important to know where they live (now and in the future), the barriers they face and the capacities they bring to change these processes of development. But people pursue different activities and practices in different places, including in rural and urban areas, which generate different environmental and socioeconomic challenges and opportunities. Furthermore, different places are characterised by varied resource endowments and by unique ecologies that emerge through adaptation to local conditions and processes of change. Again, this ensures that the nature of the challenges and opportunities for sustainable development are going to be locally distinct and, indeed, may firstly become evident at this scale.

The centrality of resources in development

Questioning environmental 'limits'

All forms of economic and social activity make demands on the resource base: as raw materials such as soil and water within agricultural production, as sources of inputs and energy into industrial production and in the construction and maintenance of human settlements and urban lifestyles. Economic development to date has been closely correlated with mounting rates of resource extraction worldwide, as shown in Figure 2.2 in terms of energy and water. Questioning such global scale human–environmental relationships and trajectories were seen in Chapter 1 to be key to understanding the requirement for sustainable development. Indeed, the predominant resource challenges for development at a global scale were considered until quite recently in terms of the limits to stocks of these natural resources such as oil, water and land, the so-called environmental 'source limits' for development. Whilst many of the predictions made in the 1970s and 1980s of oil 'running out' and water as a 'new oil crisis' (Biswas, 1993) did not materialise, the notion and timing of 'Peak Oil' remains closely debated (see Figure 2.3) and oil and water scarcity have profound

Figure 2.2a   *World energy supply by source, 1980–2009*

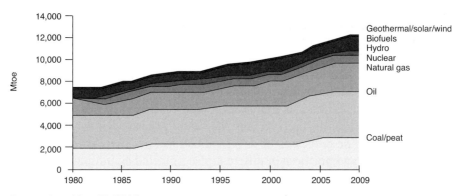

Source: adapted from IEA (2011).

Figure 2.2b   *Global water withdrawal by sector, 1900–2025*

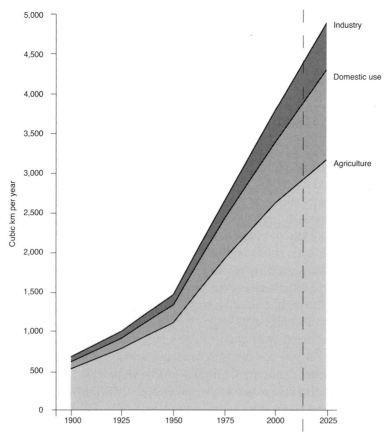

Source: adapted from UNEP (2008).

Figure 2.3 *The notion of 'Peak Oil'*

The theoretical point at which half or more of the world oil supplies are used up. Given that oil was formed through geological processes in the deep past, this means that all the oil that there will ever be in a human lifespan already exists. Once society moves past the point of Peak Oil and into using the second half of the inevitably limited supply, it is argued that oil becomes more scarce, more scattered and more costly to extract. As such the era of cheap oil and fossil fuel dependence ends.

impacts on the prospects for sustainable development globally, as seen in subsequent sections.

All production and consumption activities also produce wastes in the form of various gases, particulate matter, chemicals and solid matter. Past development processes have depended substantially on the capacity of natural systems to absorb, transport and dissipate such wastes; termed environmental 'sink' functions. Where the rate of waste generation exceeds the natural capacity of the atmosphere, oceans, vegetation or soils to absorb these, there are detrimental effects to human health and to the operation of ecological systems. In 1991, the first deaths from air pollution in Britain for more than 30 years occurred in London due to smog build-up from traffic fumes (Bown, 1994). Currently over 2 million people globally die prematurely every year through indoor and outdoor air pollution, largely in developing regions.

Finding physical space for the disposal of solid wastes is an ongoing challenge, particularly in more developed countries where rates of production are highest. In the UK, for example, the average person produces approximately 500 kg of domestic waste annually and whilst the amount that is recycled is increasing, more than two-thirds continues to go to landfill (DEFRA, 2009). New challenges for sustainable development also arise as the nature of production and consumption changes. 'E-waste' (comprising all forms of electrical and electronic equipment including PCs, refrigerators, air conditioners and mobile phones that have been discarded by their users) is now the fastest growing waste stream in industrialised countries (BAN, 2002). It is growing globally by an estimated 3.5 per cent every year (UNEP, 2007). Whilst there are opportunities in re-use and recycling of e-waste including the extraction of valuable materials such as copper and gold, e-waste also includes toxic substances of environmental health concern including lead and mercury. Box 2.1 considers some of the challenges for sustainable development presented by e-waste.

Box 2.1

E-waste: global recycling or waste trafficking?

E-waste is a complex and rapidly expanding waste stream. It also constitutes a significant business opportunity, an expanding economic sector particularly in developing countries, and is increasingly traded across international boundaries. The 'export' of e-waste is regulated under the Basel Convention on the Control of Transboundary Movement of Hazardous Waste and their Disposal because of the highly toxic elements involved including cadmium, mercury and lead that require sound environmental management if not to be hazardous to human and environmental health. However, e-waste also includes valuable substances including gold and copper that can be recovered, thereby providing a source of raw materials that are in demand and preventing further primary extractions. In emerging economies, e-waste imports can provide an affordable source of ICT equipment through product reclamation and repair, both of which are legal under the convention. There are strong social and economic arguments for trading in e-waste. As Obsibanjo and Nnorom (2007: 490) have suggested for Nigeria:

> More and more Nigerians today have access to computer facilities at home, school, business centres and Internet cafes. A great number also have access to mobile telephones and this is now playing a huge role in the development of the Nigerian economy . . . These advancements in ICT depend to a large extent on second-hand/refurbished electrical and electronic equipment.

E-waste imports can be seen as a way of supplying the rapidly increasing demand for ICT in these countries, supporting the breakdown of the global 'digital divide' and providing opportunities for social and economic development. The economic sense is also confirmed by the fact that used PCs can be bought in Nigeria at 30 per cent of the cost of new whilst demand for such 'second-hand' goods in exporting countries such as in Europe and the US is often low. The costs of recycling in exporting countries are also high, partly because of regulations on landfill and incineration of electronic goods. However, a study by the Basel Action Network (BAN) found that of over 500 container loads of second-hand PCs and accessories that are imported for reuse through the Nigerian port of Lagos every month, 75 per cent were in fact unusable, irreparable, 'e-scrap' (cited in Obsibanjo and Nnorom, 2007). BAN is one of many NGOs that are campaigning for a global system of labelling to confirm usability of second hand appliances intended for export.

In addition to this question of trade in e-waste occurring in the name of recycling but actually constituting 'dumping', there are substantial concerns regarding the environmental and health impacts during the methods and processes of reclamation and recycling and in the ultimate disposal of wastes. In many receiving countries, a lack of national regulation and/or enforcement has enabled the rapid growth of an informal sector centred on trading, repairing and recovering materials from e-waste. Whilst this

Figure 2.4 E-waste recycling sources and destinations

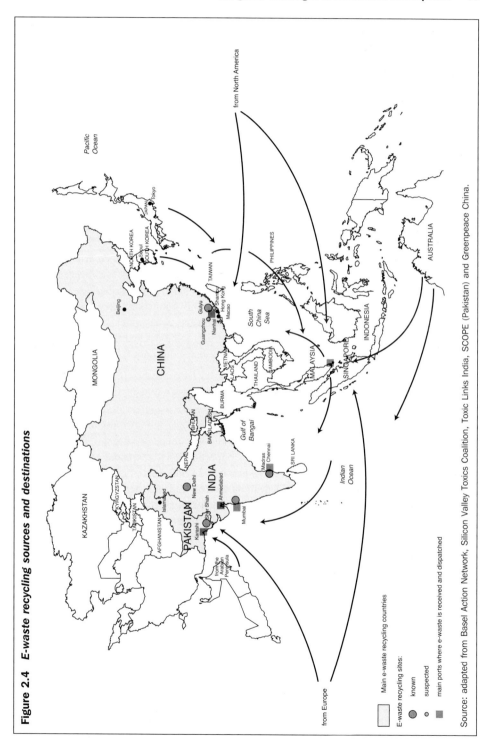

Source: adapted from Basel Action Network, Silicon Valley Toxics Coalition, Toxic Links India, SCOPE (Pakistan) and Greenpeace China.

provides opportunities for some people to gain an income (and to build a business if they have sufficient capital and networks, for example), much of the recycling activities are actually taking place within crude 'backyard operations' and at landfill sites with little formal regulation and where the health costs fall largely on the urban and rural poor. The persistence of an international trade could also mean that there is less incentive within exporting countries to invest in the infrastructure necessary and develop the market for recycling and disposal 'at home'. Figure 2.4 identifies the principal destination countries and cities for e-waste.

In Guiyu town in the southern Guandon Province of China, an estimated 100,000 people are employed in processing e-waste largely from the US, but also from Japan, South Korea and Europe. Ship loads of containers come through the port of Nanhai and then hundreds of trucks depart each day to Guiyo town (BAN, 2002). The work is based on receiving and recycling obsolete and broken computer equipment: heating printed circuit boards over charcoal burners to release possibly reusable computer chips; burning of wires in the open air to recover copper; treatment of circuit boards in open acid baths next to rivers to extract precious metals; dismantling printer cartridges to collect remnant traces of toner and for recyclable plastic and aluminium parts. Workers in these enterprises are totally unprotected: there is no basic safety equipment, they breathe in toner fumes and are exposed to heavy metal contaminants. Local water sources have been contaminated to the extent that drinking water has to be trucked in. As a method for resource/material recovery, this is also inefficient in that it does not fully release the raw materials that could be.

Whilst China is a signatory to the Basel Convention and has national legislation prohibiting imports of solid wastes that are unusable as raw materials and severely regulates imports of wastes that can be used as raw materials, the infrastructure to enforce the legislation is extremely lacking (with some additional concerns as to the political will to enforce it). The illegal nature in turn ensures opportunities for exploitation: impoverished farmers are willing to take the health risks and risks of prosecution through their desperate need for money, as evidenced by the fact that such activities often take part under the cover of darkness or re-start immediately once officials leave. Corruption is also evident as officials take money from operators in the e-waste business.

Concern for the 'sink' limits of the earth's systems also underpins the global challenge of climate change. Rising concentrations of greenhouse gases in the atmosphere due to human activities are now understood to be causing unprecedented increases in the average surface temperatures worldwide. Whilst the implications of climate warming for the functioning of various environmental systems and for societies continue to be closely researched (and debated), many earth system scientists are now exploring environmental limits in the sense of thresholds or 'tipping points'. In short, in these cases, environmental changes do not progress slowly or in a linear manner (as in the build-up of domestic wastes, for example) but may change suddenly, dramatically and possibly irreversibly.

However, what constitutes an environmental 'limit' on development is also shaped by the values and decisions of society; in determining 'acceptable' levels of resource degradation and which groups of people, living where should bear the cost of such degradation, for example. Furthermore, different interests and groups in society may value different and competing resource and environmental functions at the same time and in particular places. Reconciling these differences will be seen to underpin many of the challenges of sustainable development through this chapter. As Buckles and Rusnak (1999: 2) suggest, 'conflict over natural resources such as land, water and forests is ubiquitous'. Whilst the causes of human conflict are complex, natural resource use, by the very nature of biophysical and ecological processes, connects the actions of one individual or group to another and thereby contains the potential for conflict.

Energy futures and the challenges of low-carbon development

Energy presents two key global challenges for sustainable development: concerns for the adequacy (and security) of affordable supplies into the future and the environmental damage occurring currently through the consumption of energy. Current rates of growth in energy demand are unprecedented historically. In China alone, energy consumption grew at nearly 10 per cent per year between 2000 and 2005 and total world energy consumption is predicted to rise by 49 per cent (over 2007 levels) by 2035 (UNEP, 2007). Whilst over 80 per cent of this increase is projected to come from developing countries, the more economically advanced countries currently account for almost 50 per cent of world energy demand (UNEP, 2007). As seen in Figure 2.2a, fossil fuels such as oil, coal and gas supply over 80 per cent of the world's current energy needs. Fossil fuel use is also, however, the principal source of the global increase in carbon dioxide emissions (IPCC, 2007) making a 'global energy revolution' central to meeting the challenges of climate change (World Bank, 2010a).

The implications for sustainable development of these two issues of future energy supply and current patterns of consumption are extremely wide ranging. They include finding new stocks and alternative sources of energy and making development more energy efficient as well as less carbon intensive. They also include meeting the aspirations of those people who have historically been excluded from these opportunities. To date, access to commercial energy sources (fuels such as oil, coal, gas and electricity which have a commercial value and are often traded between countries) has been highly inequitable. The International Energy Agency has estimated that the 20 million inhabitants of New York State use the same amount of electricity as 800 million people across sub-Saharan Africa (Webb, 2011). In the mid-1990s, it was estimated that the annual commute by car into New York City alone used more oil than the whole of Africa (excluding South Africa) in one year (Edge and Tovey, 1995: 319). Figure 2.5 confirms the stark regional differences currently in energy use per capita worldwide. In many developing countries, as much as 90 per cent of total energy consumption may be supplied by biomass sources such as fuelwood, charcoal and animal dung (that are increasingly traded, but within informal rather than formal markets).

The environmental and social sustainability of the continued dependence on oil globally is the focus of close academic and political debate. In Box 2.2, different understandings of tar sands in Canada as a future global energy source or a 'curse' threatening

Figure 2.5 *Total primary energy supply per capita 2007, selected regions and countries*

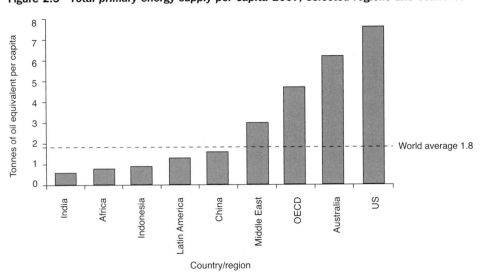

Source: OECD (2010d).

Canada's previous good environmental record is highlighted. Box 2.3 considers the risks and opportunities of new sources of energy based on biofuels. The capacity of further technological advancement in developing new sources of oil was also seriously questioned through 2010 subsequent to the 'BP disaster' in the Gulf of Mexico. The initial explosion on the deep water drilling platform killed 11 people and had widespread environmental, social and political ramifications across the region, for the local people, the company and for the administration of President Obama. Internationally, countries were prompted by the disaster to consider their own energy independence and commitments to renewable energies.

Box 2.2

Tar sands in Canada

'Tar sands' are deposits of bitumen mixed with sand and clay. They are considered a very 'inferior' source of hydrocarbon in that they contain large amounts of carbon and very little hydrogen, require a great deal of energy and water to process into usable oil and produce three to five times as much greenhouse gas as conventional oil on extraction. Canada is currently producing 3 million barrels of oil a day from tar sands covering 140,000 km², largely in Alberta. Canada's rapid development of tar sands (from 600,000 barrels a day in 2000 to 1.3 million barrels currently, with a planned increase to 3.5 million by 2030) are argued to be a 'curse' for the country. Canada is the only signatory to Kyoto that has abandoned its targets and is one of world's top ten greenhouse gas emitters. However, it was once celebrated for its environmental record and pristine forests in particular. The exploitation of tar sands threatens ancient boreal forest and freshwater systems, with implications for bird migrations and wildlife habitats. Much of the oil extraction also occurs on lands belonging to First Nation Communities. Yet these are projects in which international banks, pension funds and corporations (including BP) are investing. The capacity of Canada's national government to direct energy policy (and to restrict tar sands development) is also constrained by the North American Free Trade Agreement (NAFTA) that deregulated the oil sector, removed restrictions on foreign ownership and prevented import/export restrictions. Furthermore, in the context of rising oil prices and the US's search for 'secure' energy sources, there are signs that Canada is becoming an 'increasingly aggressive defender of hydrocarbons' (Nikiforuk, 2010: 8).

Source: compiled from Nikiforuk (2010) and Worth (2010b).

Box 2.3

The risks and opportunities of biofuels

In 2008, the risks and opportunities of biofuel production were the focus of the Food and Agriculture Organisation's *State of Food and Agriculture Report*. At the time, there were high expectations concerning liquid biofuels and the potential to mitigate climate change, contribute to energy security and support agricultural producers around the world. Many governments were introducing policies and measures to enhance biofuel production justified in these terms. However, there were also mounting concerns for food security, particularly for the world's most vulnerable people. Food prices were higher in real terms than at any time since the 1970s and were rising rapidly, prompting civil unrest and protest in many countries. A key concern for many UN agencies, donors and NGOs was the contribution of the recent growth in the use of agricultural commodities (maize, sugar, oil seeds and palm oils) and land for the production of biofuels to this emerging food crisis.

The report concluded that biofuels were having minimal influence on world energy use (and greenhouse gas emissions) but substantial impact on agriculture and food security. Liquid biofuels remain a very limited part of total energy supply and even in the transport sector, delivered only 0.9 per cent of total fuel consumption in 2005 for example (IEA, 2007). There is also concern that any emissions 'savings' from such use is more than countered by the increased emissions that occur through the land use changes required to produce those crops. Whilst the impact on carbon emissions depends on the particular crop, where and how production occurs, the report identified that biofuel production strongly enhances the risk of large-scale land-use change including deforestation and in many cases, the net effect on emissions is unfavourable. Further environmental impacts include depletion of water resources and biodiversity.

The report identified the fact that the rapidly growing demand for biofuel feedstocks had contributed to higher food prices and undermined food security of the poor in both urban and rural areas. Demand for agricultural feedstocks for liquid biofuels pushes up prices of these goods and the resources (fertilisers etc.) used to produce them, which impacts most heavily on the poor. Poorer households spend a larger proportion of their income on food than richer households and even in rural areas, the poor are net purchasers of food. Whilst food prices are influenced by diverse factors including weather-related production shortfalls, increasing fuel costs and population growth and urbanisation, the report urges carefully targeted safety nets to ensure access to adequate food on behalf of the most vulnerable.

To date, the production of biofuels has principally been in the US and EU. However, high targets are being set, particularly in South East Asia where investment is largely by overseas companies and for export (DEFRA, 2008). For example, Indonesia has already cleared 18 million hectares of forests for palm oil production and an additional 20 million hectares is identified in regional development plans for further expansion.

Whilst there are now international standards (such as the Roundtable on Sustainable Palm Oil) developed by the industry in conjunction with conservation organisations and social justice groups (and Indonesia has its own laws protecting customary rights), there remain concerns that legal reforms will take some time to be effected and much of the expansion of biofuels will continue to be at the expense of existing tropical rainforest and local livelihoods.

If future demands for energy are to be met and the emissions reductions required to address climate change are to be achieved, there is widespread understanding that lower carbon growth is an essential global challenge. Lower carbon growth includes raising energy efficiency in production, that is, reducing the amount of energy used to produce each unit of output. It also involves reducing the amount of carbon emitted for a given amount of energy used as different fuels produce varied levels of carbon (coal being much more carbon intensive than gas, for example). A widely used measure of the carbon intensity of society is the carbon dioxide produced to deliver a 'thousand dollars of Gross Domestic Product'. Table 2.1 confirms that the carbon intensity of the world economy has been declining in recent years. It also shows that economic production in the rapidly developing economy of Brazil is currently less carbon intensive than in the US or UK.

There is optimism that developing countries can achieve faster rates of improvement in carbon intensity of production through their industrial transformations than occurred in North America and Europe based on the more energy efficient technologies that are now

Table 2.1 *Carbon intensity of GDP*

	Metric tons carbon dioxide per US$1,000 GDP			
	1990	*1995*	*2000*	*2005*
World	0.64	0.63	0.56	0.56
US	0.71	0.66	0.60	0.55
UK	0.48	0.41	0.35	0.31
China	2.03	1.44	0.98	1.13
India	0.65	0.76	0.64	0.55
South Africa	1.72	1.93	1.87	1.66
Brazil	0.26	0.27	0.29	0.27
Russia	–	1.59	1.41	1.13

Source: compiled from Energy Information Administration (2006).

available. Forty per cent of energy use in Brazil, for example, comes from renewable energy sources (Flavin, 2008). Whilst China remains heavily dependent on carbon-intensive coal sources for energy production, it has set ambitious targets on renewables in recent years and is achieving annual improvements in energy efficiency and in lowering the carbon intensity of production (IEA, 2009). However, there are no simple linkages between levels of development and carbon intensity. Energy use is influenced by the structure of economy (manufacturing and mining are more energy intensive than agriculture, for example) as well as by climate, which influences energy use such as for domestic heating and cooling. Energy policies are also key: countries with higher energy prices and more stringent regulations tend to be more energy efficient (World Bank, 2010a).

Evidently the challenges of lower carbon growth include enhancing the use of renewables (such as solar, wind, biofuels, nuclear and hydro-electric sources) that provide low to zero carbon emissions. Several developing nations are now world leaders in the use of renewable energy, suggesting further optimism for lower carbon growth in future. Brazil is currently the world's largest producer and consumer of renewable energy and India has the fourth largest wind power industry worldwide (Flavin, 2008). Investments in renewable energy have been rising globally with many countries now having national commitments and targets on renewable energy, often closely linked to climate change actions (see Chapter 3). Such patterns and the substantial known potential of technologies and policies for further energy efficiencies have led to the suggestion that developing nations can 'leap-frog' more developed nations in their energy trajectories and lead the way towards lower carbon futures. However, 'the fact that so much efficiency potential remains untapped suggests that it is not easily realized' (World Bank, 2010a: 190). There is also concern as to whether such 'decarbonisation' of development trajectories can occur at a rate fast enough to stabilise climate warming at an acceptable level, as discussed below.

Recent efforts to enhance the use of renewable energy sources have confirmed that the challenges of sustainable development encompass difficult political choices and trade-offs concerning sustainability objectives. Renewable energy technologies are expensive to 'get started' and put into operation, at a scale that is understood as essential to substantially influence national and global energy futures. Such investments have often relied on political will in the form of government subsidies and have been closely influenced by general economic conditions (including oil prices) to make them economically viable. Such circumstances can change quickly, as seen in late 2008 when investments in 'clean energy' worldwide fell by over 50 per cent compared to 2007 levels as the global economic crises emerged (World Bank, 2010a). Many political leaders were confronted very directly with the challenge of sustainable development that often include prioritising decisions for the longer term (and for 'global' benefit) over more immediate and overtly pressing domestic issues. Furthermore, the biggest factor explaining public opposition to wind technology in Europe is concern over the visual impact of wind farms on landscape values (Wolsink, 2007), confirming that environmental resources and functions can be valued in very different ways by different 'interests' and reconciling these are key challenges for sustainable development.

Finding alternatives to fossil fuels within the transport sector is particularly important for moving to lower carbon development patterns. This sector accounts for 30 per cent of total global energy demand and is responsible for 21 per cent of global annual greenhouse gas emissions (DEFRA, 2008). However, whilst 'biofuels' based on maize, sugar, oil seeds and palm oils, for example, offer potential for emissions savings, Box 2.3 confirms how they may compromise other sustainability objectives such as food security at both global and household levels.

Water 'scarcity'

The close relationship between past development and rising demands for water was seen in Figure 2.2b and is the basis of long-standing sustainability concerns over water availability and of securing future sources to support development. In the early 1990s, the fears were that increasing physical scarcity of water could lead to 'water wars'. Over 60 per cent of the world's freshwater supply lies in over 260 international river basins, spanning borders of more than two countries (Clarke and King, 2004). Two in every five people globally live in such basins (UNDP, 2006: 20). The concern was that competition over water could threaten already fragile ties between states in key regions like the Middle East. In 2005, water scarcity was identified as a 'globally significant and accelerating condition' (MEA, 2005: 51). Global water use relative to accessible supply increased by 20 per cent per decade from 1960 to 2000 (ibid.). It is predicted that water withdrawals in developing countries will increase by a further 50 per cent to 2025 (UNEP, 2007). By the same date, it is estimated that 1.8 billion people worldwide may live in water-scarce environments as regional and local precipitation patterns shift under climate change (ibid.). A key sustainability concern is that reductions in precipitation will impact hardest in regions already affected by drought, such as sub-Saharan Africa, that also have the lowest levels of human well-being (MEA, 2005). Many of the predicted impacts of climate warming are water-related (discussed further below) including the disruption of ocean currents and increased flood events. In the last decade, the devastating tsunamis in Thailand (2004) and Japan (2011), Hurricane Katrina in New Orleans (2005) and record-breaking floods in, among others, the UK, Australia, the US and Pakistan have all brought water issues prominently into people's environmental consciousness worldwide.

Water scarcity currently presents a number of complex and interrelated global challenges for sustainable development. Whilst water wars have generally been avoided through cooperative solutions and treaties, many international agreements in water-stressed regions continue to be tested. Box 2.4 identifies rising tensions with India over the Indus Water Treaty as one of the current challenges around water for Pakistan. The box also highlights how declining water availability can be a source of conflict and unrest at a local scale, for example when farming livelihoods become untenable and as competition from urban domestic demand and industrial uses rise. However, it is the agricultural sector globally that uses the majority of freshwater resources (see Figure 2.2b) and further increases in food production will be needed to feed expanding populations and address undernourishment and food insecurity (see Young, 2011 in this series). Managing water as a productive resource for livelihoods efficiently and equitably is an immense challenge facing governments worldwide (UNDP, 2006, 2007).

The global challenge of ensuring greater equity in access to water as a fundamental foundation for human development has also been recognised within the Millennium Development Goals. Currently over 1 billion people worldwide do not have access to safe drinking water and 2.6 billion lack basic sanitation and safe removal of waste water. Whilst it is anticipated that the target to halve the proportion of the people unserved by 2015 may be met in respect to drinking water coverage, there remain substantial rural–urban and wealth disparities. The goal to halve the proportion of people without access to improved sanitation and waste disposal is not expected to be met. These challenges are considered further in Chapter 5.

Box 2.4

Complex concerns around water in Pakistan

Despite the national and global attention that is given to the country's troubles with terrorism, insurgency and a fragile political order, it may be Pakistan's little-noticed water crisis that costs the greatest number of lives. In the long term, this may prove its most destabilising political issue.

(Waraich, 2010: 34)

In July 2010, record breaking monsoon rainfall in Pakistan led to over 1,000 deaths and 1 million people affected through flooding. Yet, many more perish daily from water-related diseases (630 children a day from diarrhoea, for example) and water availability per person per year has fallen in the country by 80 per cent in the last 60 years. One factor has been population growth; Pakistan is currently the sixth most populous country in the world with a population of 175 million that is growing at 1.6 per cent per annum. In addition, much of Pakistan's water comes from glaciers that are shrinking with climate change. Water supply in Pakistan is also heavily dependent on the flows of six rivers from Indian-controlled Kashmir and the Indian Punjab. Whilst these rivers are the subject of the 1960 Indus Water Treaty, the treaty is under strain and was an issue that featured in the recent failed peace talks between the two countries. India is currently constructing a water diversion scheme in Kashmir that could prevent water from the Jhelum river reaching Pakistan, and also has more than 20 hydro projects active on three western rivers (allocated to Pakistan under the treaty). India argues that these hydro schemes will return water downstream and therefore don't violate the treaty.

Ninety per cent of water use in Pakistan is for agriculture, which employs approximately half of the national workforce and accounts for a quarter of exports. Declining water supplies are already affecting production in the Punjab (the country's 'bread basket'), where small scale producers have had livelihoods ruined and there are concerns of rising local unrest that could provide a focus for militant activity. Whilst water- and sanitation-related diseases are estimated to cost Pakistan's national economy nearly £1 billion a year, investments in water infrastructure, in treatment technologies and in improving agricultural techniques to reduce waste and enhance efficiency, for example, have yet to be implemented.

Source: compiled from Waraich (2010).

Some commentators continue to argue that the roots of these challenges are in the rising demands of an expanding population on the physical water resources of the globe. However, it is more widely understood that 'the scarcity at the heart of the global water crisis is rooted in power, poverty and inequality, not in physical availability'

(UNDP, 2006: 2). In water stressed parts of India, for example, wealthy farmers are able to pump water from aquifers 24 hours a day, whilst poorer, neighbouring farmers rely on rain fed agriculture (ibid.). In the West Bank, Israel has restricted Arab farmers from drilling new wells (leading to decline in irrigated area, loss of productivity and increased unemployment) whilst Israeli settlers are able to continue to dig deeper wells, in some cases causing water tables to fall beyond the reach of the Palestinian wells (Mastny and Cincotta, 2005). Figure 2.6 illustrates what is termed a 'perverse principle' that applies widely across the developing world in terms of accessing domestic water: that the poorest people not only get access to less clean water, but they also pay higher prices (UNDP, 2006: 7). Whilst there is undoubtedly a substantial challenge for particular regions for delivering on the targets for clean water and sanitation, domestic water use represents a very small fraction of national water use and there are many illustrations of inequalities in access being the product of political choices rather than water availability (discussed further in Chapter 5). Water shortages for agricultural development at a country level are also understood to be one factor driving the recent pattern of large scale leasing of land overseas (particularly within Africa) to deliver domestic food requirements. One recent case is of a Saudi billionaire businessman who has purchased 1,000 hectares of land on a 99-year lease in Ethiopia to grow wheat, rice and vegetables for the Saudi market (Vidal, 2010). A key sustainability concern of these and related land transfers (as seen in Boxes 2.3 and 2.4) is that people with the weakest rights are often those that lose access to land and water when more powerful constituencies become involved (Cotula et al., 2009).

Figure 2.6 *The 'perverse principle' in accessing clean water*

- People living in the slums of Jakarta, Indonesia; Manila, the Philippines; and Nairobi, Kenya, pay 5–10 times more for water per unit than those in high-income areas of these cities – and more than consumers pay in London or New York.
- High-income households use far more water than poor households. In Dar es Salam, Tanzania, and Mumbai, India, per capita water use is 15 times higher in high-income suburbs linked to the utility than in slum areas.
- Inequitable water pricing has perverse consequences for household poverty. The poorest 20 per cent of households in El Salvador, Jamaica and Nicaragua spend on average more than 10 per cent of their income on water. In the UK, a 3 per cent threshold is considered an indicator of hardship.

Source: compiled from UNDP (2006).

The resource curse in development

Whilst resource scarcity and degradation are evidently global challenges for sustainable development, there is also concern as to the relationship between natural resource abundance and the prospects for sustainable development. Auty (1993) was one of the first to suggest a 'resource curse' on the basis of his research that found slower rates of economic growth in 'mineral rich' economies (defined as having more than 40 per cent of exports based on hard minerals such as copper, tin and bauxite) than in non-mineral economies. In 2004, Christian Aid compared six oil producing nations (Angola, Iraq, Kazakstan, Nigeria, Sudan and Venezuela), with six non-oil-producers (Bangladesh, Bolivia, Cambodia, Ethiopia, Peru and Tanzania) over four decades in terms of their human development performance. They found that the oil economies had significantly slower economic growth, and life expectancy and literacy rates improved more in non-oil economies than oil economies.

Whilst the precise links between natural resource endowments and human development continue to be researched, there is increasing concern for how favourable natural resource wealth can be a source of violent conflict and host to conditions for criminal and terrorist activities. This has become particularly evident since the '9/11' attacks on the World Trade Center in New York in 2001 and the subsequent 'war on terror', but is also evident in the global concern for future energy/oil security. In countries such as the Congo and Sudan, whilst the trigger for civil unrest and conflict may not have been the minerals or oil per se, it is understood that the pillaging of those resources generates the finances that allow conflict to continue and make it harder to resolve. Often, these lucrative resources represent one of the few sources of wealth in otherwise very poor societies and conflicts are very much centred on gaining and maintaining control over them. It is also understood that conflict creates territories outside the control of a recognised government which often become havens for terrorism, presenting global challenges as confirmed by the current situations in Pakistan and Afghanistan.

There is also widening concern that resource extraction in many parts of the developing world, whilst not fuelling full-fledged civil (or international) wars, is having profound consequences. In many instances, the benefits of logging, mining and so on go to government elites and foreign investors, whereas the burdens are felt by local people (and more insidiously) in terms of loss of land, environmental devastation, social impacts and the abuse of human rights. One of the most widely known examples is in the Niger delta of Nigeria (see Adams, 2009). Since the 1970s, oil corporations (most notably, Shell) and the government have received the benefits of oil extraction and the Ogoni people have suffered degradation of the resources on which their livelihoods depend (such as through the loss of land directly and the sterilisation of soils and water pollution) and are living in increasingly dangerous and unhealthy environments. Mass protests by the Movement for the Survival of the Ogoni People (MOSOP) in the 1990s were met with violence and a state campaign of encouragement of ethnic violence. Over 2,000 Ogoni people were killed, including Ken Saro-Wiwa (the internationally known spokesman for MOSOP) and eight other leaders who were executed by the government itself. Whilst the oil industry is no longer active in the area, a recent environmental assessment of Ogoniland (UNEP, 2011) confirmed widespread oil contamination throughout the region that was severely impacting on many components of the environment and human health. The report concluded that the operation and maintenance (and the decommissioning and remediation) of the oilfield had been inadequate and full environmental restoration may take 30 years. However, it also urges immediate, emergency measures to address the human health impacts of widespread water contamination in particular and suggests that these should be funded by the oil industry and the government.

Beyond these more obvious, high-profile resource conflicts, many more conflicts occur at local levels centred around access to basic resources for livelihood. Too often, development patterns in the past have served to remove, or compromise in some way, the opportunities for some groups of people over others. The intensity and level of resource conflict varies hugely, as do the factors underpinning them; indeed, many conflicts have multiple causes occurring as resources become more scarce, as new interests enter an area and raise competition over resources or where particular ethnic identities or ways of life that are linked to those resources become threatened (or through any combination of these and other dimensions). Whether political, class or social and cultural dimensions dominate:

> power differences between groups *(of resource users)* can be enormous and the stakes a matter of survival. The resulting conflicts often lead to chaotic and wasteful deployment of human capacities and the depletion

of the very natural resources on which livelihoods, economies, and societies are based. They may also lead to bloodshed . . . As in other fields with political dimensions, those actors with the greatest access to power are also best able to control and influence natural resource decisions in their favour.

(Buckles and Rusnak, 1999: 3)

Questions of population, resource demand and consumption

Questions of population are central to the global challenges of sustainable development in that the number of people present in the world is one factor that 'scales' humanity's impact on the planet (Jackson, 2008). Questions concerning the relationship between population, resources and development were seen to be central in the emergence and shaping of mainstream discourses of sustainable development in Chapter 1. For example, the 1970s was an era of rapid population growth globally, and particularly in the developing world. Influential 'global future scenarios' including *The Limits to Growth* (Meadows et al., 1972) and *The Population Bomb* (Ehrlich, 1968) argued for preventative population control in developing countries. In contrast, a very different view of the population–development relationship was articulated at the Stockholm conference in 1972 where representatives of the developing nations argued that population was not the problem, rather the lack of development. Their view was that 'development was the best form of contraception'.

Between the time of the WCED reporting in 1987 and 2007, global population increased from 5 billion to 6.7 billion at an average rate of 1.4 per cent per annum (UNEP, 2007). At the time of writing, the number of people worldwide will imminently exceed 7 million. By 2050, it is projected to be over 9 billion (UNPD, 2009). Environmentalists have been accused of not facing up to the questions of population growth, among other things because of the highly contentious, political and personal issues embraced in questions of population planning. In recent years, a number of high-profile environmentalists such as David Attenborough have spoken out:

> I've never seen a problem that wouldn't be easier to solve with fewer people, or harder, and ultimately impossible, with more. Population is reaching its optimum and the world cannot hold an infinite number of people.

(Attenborough cited in Bahra, 2009)

However, the drivers and characteristics of population change are complex (see Gould, 2009 in this series) and projections of future patterns are difficult. The UN Population Division (widely cited as the most authoritative source of statistics) models a number of 'variants' within their projections. These encompass different assumptions regarding the components of population change (mortality, fertility and distribution). Under their 'medium' variant, the majority of future population growth will be in less developed regions where 86 per cent of the world's population is predicted to live by 2050 (Figure 2.7), although the rate of growth in all regions is projected to slow, as seen in Figure 2.8. Such projections assume all countries will gradually move through similar demographic transitions; no account is taken of potentially different policies, responses and development trajectories in particular countries. Yet, as Gould suggests, 'China is not likely to follow the same development path as Chad' (2009: 271). Furthermore, as seen through this text, more sustainable development trajectories are going to depend on substantially different policies, technologies and governance arrangements, for example, than have been seen to date. The particular challenges of population for sustainable development will therefore be shaped at these scales.

There is no doubt that processes of population change in large and populous nations such as China and India have had (and will continue to have) an impact on global patterns. Approximately 37 per cent of the world's population lives in China and India (UNPD, 2009). Rapid economic development in these countries in recent years was seen in the section above to have substantially changed the global energy system. In the following section, the achievements made in these countries in moving very large numbers of people out of poverty will be seen. Combined, these patterns have implications globally for rising consumption and resource demands into the future.

Issues of population also shape global challenges of sustainable development in terms of where people live, what barriers they face and what capacities they bring. The largest form of population redistribution currently is urbanisation. Fifty per cent of the world's population now lives in towns and cities. Between 1995 and 2005, the urban population of the developing world grew by an average of 1.2 million per week or 165,000 per day (UNHSP, 2010). However, there is substantial variation between and within countries as to the pace of urbanisation and the growth rates of particular cities, for

Figure 2.7 *Distribution of world population (percentage) by world region*

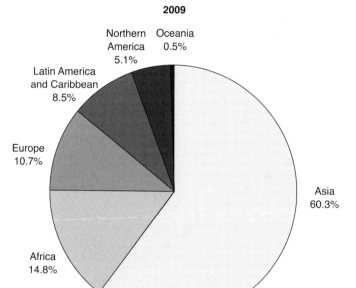

2009

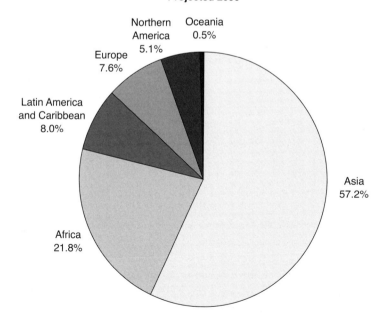

Projected 2050

Source: UNPD (2009).

Figure 2.8 *Projected average annual rate of population change by major regions, 2010–2100 (medium variant)*

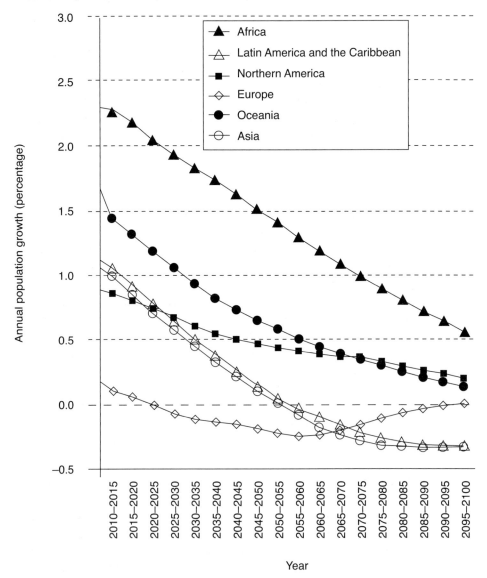

Source: adapted from UN-DESA (2011) World Population Prospects: the 2010 Revision, New York.

example. The rising numbers of people resident in urban areas brings a range of challenges and opportunities for sustainable development (considered further in Chapter 5) that include meeting the needs for housing, employment and infrastructure that can be particularly difficult in rapidly expanding cities. Yet, cities are also major drivers of development and are a source of political momentum that can advance actions towards sustainability. World cities consume more than two-thirds of global energy and produce more than 70 per cent of carbon dioxide emissions (World Bank, 2010a: 194). Yet much of the energy consumed within cities is used in the production of goods that are consumed elsewhere.

The numbers of people migrating internationally has also increased in recent decades doubling from 1960 to 190.6 million in 2005 (Williams et al., 2009: 159). However, most international migrations are either between developing countries or between developed countries. International migration in relation to total world population has also remained relatively stable over the last 50 years at 3 per cent (UNDP, 2009). The flow of money from migrants to family back home ('remittances') is also a very significant source of finance for development. At a global level, it is estimated to be over 2.5 times the value of official aid finances (UNDP, 2009). In Chapter 4, the wider value of remittances for individual households in terms of social support and knowledge transfers is considered. Furthermore, migration at a variety of spatial and temporal scales is now understood as an important strategy through which people are able to reduce their vulnerability to both environmental and non-environmental risks and to accumulate assets and move forward in their lives (Tacoli, 2009).

However, an estimated 42 million people currently are displaced from their homes by conflict or persecution, with 15.2 million people living outside their country of origin as refugees (UN-DESA, 2010: 15). Such involuntary migration also presents profound global challenges for sustainable development that are beyond the scope of this text. However, it is anticipated that they may become worse under climate change. Whilst predictions are difficult due to uncertainty regarding the effects of climate change but also the complexity and diversity of reasons why people migrate or seek refuge, Christian Aid (2007) suggests that there may be 1 billion forced migrations by 2030 as climate change exacerbates existing problems. It is unlikely that a 'climate refugee' will be easily or suddenly spotted, rather gradual changes in climate combined with other political, economic and social factors will drive changing patterns and processes of migration (Forum for the Future, 2010).

Evidently, questions of population numbers, distribution and location are important in understanding the global challenges of sustainable development. However, factors of 'how we live' (levels of consumption and lifestyle expectations at individual and societal levels) are also profoundly important in shaping humanity's impact on the planet. 'Ecological Footprinting' is a way of quantifying how much particular citizens extract from the planet to live the way they do (discussed further in Chapter 6). It is calculated that if everyone consumed as much as the average UK citizen, it would need more than 'three Earths' to support them (NEF, 2009). There are deep challenges for sustainable development associated with changing patterns of consumption, particularly in higher-income countries, considered further in Chapters 3 and 6. However, it is the *inequalities* in 'how we live' globally that are increasingly understood as central to the challenges of sustainable development. As identified by the WCED (1987: 95):

> threats to the sustainable use of resources come as much from inequalities in people's access to resources and from the ways in which they use them as from sheer numbers of people.

But inequality within and between countries is rising globally (Milanovic, 2005). Tackling inequality within current generations is at the centre of concerns for sustainable development. It affects the prospects for global society as a whole, not just for the least well off. In an extensive analysis of a range of social and health impacts of inequality in more advanced economies, Wilkinson and Picket (2009) have shown how social mobility and trust as well as literacy, life expectancy and child well-being are all much higher in more equal societies. Incidence of mental illness, obesity and homicide rates, for example, are much worse in less equitable ones.

It will be seen regularly through this text that persistent (and often deepening) inequalities underpin many of the challenges of and actions being undertaken towards sustainable development. These embrace inequality not only in terms of income, aspects of consumption and contribution to environmental degradation, but also in terms of opportunity and power, ranging from access to resources and environmental improvements (such as energy, water and sanitation highlighted above) through to having freedom of choice and action to shape a life you value (Sen, 2000; MEA, 2005). Figure 2.9 illustrates how opportunity to benefit from a consistent education is shaped by, among other things, wealth, location and gender. Based on data from 42 developing countries it confirms that inequalities in wealth are important in shaping the chance of children gaining an education, with fewer children out of school amongst wealthier households. However, rural children are twice as likely to be out

Figure 2.9 *Inequality in access to education 2008*

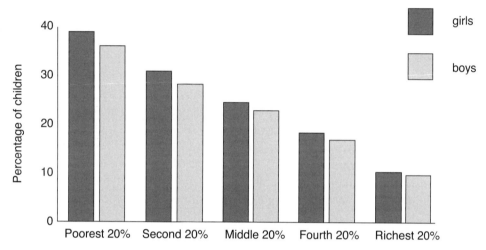

a) Out of school children by wealth and gender
Source: adapted from UN-DESA (2010).

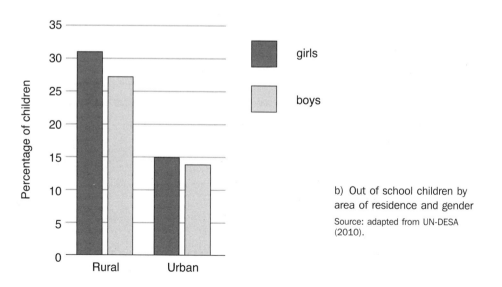

b) Out of school children by area of residence and gender
Source: adapted from UN-DESA (2010).

of school as children living in urban areas and it is girls in the poorest 20 per cent of households who have the least chance of getting an education (being four times more likely to be out of school as boys in the richest households). Whilst there are multiple reasons why children remain out of school (evidently including cost), in many countries educating girls is widely perceived as being of less value than educating boys (UN-DESA, 2010). Indeed, among the most persistent inequalities worldwide are gender inequalities. Globally, there are an estimated 100 million 'missing women', women who would be alive if it were not for practices of infanticide, neglect and sex-selective abortion (UNDP, 2002). Momsen (2010 in this series) explores complex factors including the social status of women, the economic and cultural basis of son preference and women's access to education that underpin these practices. Through this text, it will be seen how inequalities by gender cut across wealth to shape many opportunities and challenges for sustainable development. As Satterthwaite et al. (1996: 16) highlighted, 'women regularly own a very small proportion of the natural resources and often face discrimination, when compared with men, in obtaining land, education, employment and housing'. Most fundamentally, patterns of inequality confine large numbers of people to poverty. Poverty presents a persistent global challenge for sustainable development in threatening both opportunities for development and the environmental basis for livelihoods, as seen in the following section.

The challenges of poverty and inequality

Addressing the global challenge of poverty was central to the recommendations of the WCED in 1987. Meeting the needs of the poorest sectors of society was considered essential to both protect environments and ensure development:

> Poverty is a major cause and effect of global environmental problems. It is therefore futile to attempt to deal with environmental problems without a broader perspective that encompasses the factors underlying world poverty and international inequality.
>
> (WCED, 1987: 3)

For authors such as Anand and Sen (2000: 2030), a focus on inequalities within the current generation ('intra-generational equity') is central to the entire approach of sustainable development and raises deep moral questions:

> this goal of sustainability . . . would make little sense if the present life opportunities that are to be 'sustained' in the future were miserable and indigent. Sustaining deprivation cannot be our goal, nor should we deny the less privileged today the attention that we bestow on generations in the future.

Through this section, it will be seen that poverty and inequality continue to deny millions of people basic human rights in the short term and the potential to achieve development aspirations in the future. A stark illustration of the failings of current development to meet the needs of current generations is that over 24,000 children die every day before their fifth birthday, largely in the developing world and principally from preventable diseases closely associated with environmental conditions such as water quality and under-nutrition (UNICEF, 2009). Over half of these 9 million annual deaths are in Africa. However, it will also be seen that the linkages between poverty and the environment are complex and rarely direct; there are many dimensions to poverty and inequality and multiple factors that influence behaviour, including as regards environmental resource management. The section also highlights recent work that considers poverty less in the sense of 'unmet needs' and more in terms of the capacities that people have to shape their lives and take the opportunities (including those linked to environmental systems and services) for moving out of poverty.

Poverty and the MDGs

Addressing poverty and inequality were central to the commitments made in 2000 by the UN and the global community within the Millennium Declaration, as seen in Figure 2.10. With less than five years remaining to the target date for the Millennium Development Goals, it is anticipated that the first goal (MDG1) of halving the proportion of people in extreme poverty (over 1990 levels) at a global scale could be achieved (UN-DESA, 2010). The proportion of people in developing countries living in poverty (defined as living on less than US$1.25 a day) fell from 46 per cent in 1990 to 27 per cent in 2005 and is predicted to continue to fall to 15 per cent by 2015 (World Bank, 2010a: 39). Between 1990 and 2005, 400 million people escaped extreme poverty, as seen in Figure 2.11. However, there is a geography to persistent poverty: whilst some regions and countries (notably East Asia and China) have shown significant improvements, others are unlikely to meet MDG1. In sub-Saharan Africa, for example, the proportion of people in poverty was 58 per cent in 1990 and 51 per cent in 2005. Whilst India is expected to achieve the Goal (reducing poverty rates from 51 per cent in 1990 to an anticipated 24 per cent in 2015) other countries in southern Asia are not.

Figure 2.10 *The Millennium Declaration commitments to the challenge of global poverty and inequality for sustainable development*

'We have a collective responsibility to uphold the principles of human dignity, equality and equity at the global level. As leaders we have a duty therefore to all the world's people, especially the most vulnerable and, in particular, the children of the world, to whom the future belongs.'

Millennium Declaration statement 2

'We will spare no effort to free our fellow men, women and children from the abject and dehumanizing conditions of extreme poverty, to which more than a billion of them are currently subjected. We are committed to making the right to development a reality for everyone and to freeing the entire human race from want.'

Millennium Declaration statement 11

'We reaffirm our support for the principles of sustainable development, including those set out in Agenda 21, agreed upon at the United Nations Conference on Environment and Development.'

Millennium Declaration statement 22

Figure 2.11 *The number of people living on less than $1.25 a day*

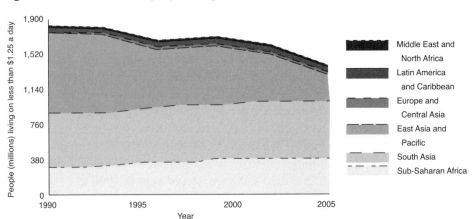

Source: World Bank Global Monitoring Information System.

Table 2.2 Child mortality rates by wealth, selected countries (under-five mortality rates per 1,000 live births)

Country	Poorest 20%	Richest 20%
Brazil	99	33
Peru	63	11
Philippines	66	21
South Africa	87	22
Nicaragua	64	19
Namibia	55	31
India	141	46
Kenya	149	91
Nigeria	257	79
Bolivia	105	32

Source: UNDP (2007).

There also remain substantial inequalities globally in the opportunity to build healthy and long lives. A child born in Norway today can expect to live to beyond the age of 80, whereas in Zambia, life expectancy is currently below 40 years and the gap between wealth and the very poorest countries is increasing (UNDP, 2009). Life expectancy in 14 African countries was lower in 2000–5 than it was in 1970–75 (UNDP, 2007). Wealth inequalities within countries also shape these basic opportunities to live a long and healthy life. In the UK, for example, the difference in life expectancy between 'professional' and 'unskilled' social classes are 7.3 years in the case of men and 7 for women (ONS, 2007). Table 2.2 confirms the large gaps that exist between richer and poorer groups within countries in terms of child survival.

Whilst differences in wealth and income inequality are evidently important in explaining spatial patterns of child mortality across scales, factors of the environment are also key. For example, 90 per cent of all child deaths due to malaria occur in Africa (UNICEF, 2010). However, more children die before the age of five in West Africa than in East Africa, linked to the environmental conditions and the disease ecology of malaria (Hill, 1991). Research has also shown that children born in drought years in East Africa were more likely to suffer stunted growth (and additional afflictions that don't go away when the rain comes) that impact on nutritional status, the likelihood of attending primary school and even their ability to bear children in the future (UNDP, 2007). A key challenge for sustainable development is that many of the main killers of children, such as malaria and diarrhoea (see Figure 2.12), are particularly sensitive to changes in ecosystems and the abundance of human pathogens, making children particularly vulnerable to the effects of climate warming (UNICEF, 2009).

Children are amongst the poorest groups in all societies. They are also responsible for very little pollution but are themselves extremely

Figure 2.12 *Major causes of death in neonates and children under 5 globally*

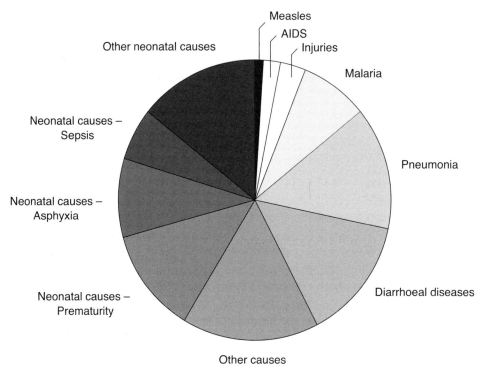

Source: adapted from UN-DESA (2010).

vulnerable to the pollution effects caused by others. In terms of their physiological immaturity for example, a resting three-year-old consumes twice as much oxygen and therefore twice the pollution weight for weight as a resting adult (Satterthwaite et al., 1996). In addition, children's underdeveloped kidneys, livers and enzyme systems are less able to process such pollutants. Children's activities may also place them at particular risk: babies instinctively suck much of what they pick up in their hands and young children play in dangerous places such as streets or waste tips, aggravating the risk of contact with contaminated sources. The persistent challenge of premature deaths of children under five confirms the unsustainable and unjust nature of current patterns and processes of development that are evidently not meeting the needs of current generations. It is not anticipated that the MDG of reducing under-five mortality by two-thirds by 2015 will be met (UN-DESA, 2010).

Defining poverty

'Poverty' is a term open to many different definitions and interpretations. The earliest and most widespread measure used to quantify 'development' achievements across a population has been Gross National Product (GNP) per capita (as compiled by the World Bank, for example). GNP refers to the total value of all the goods and services produced and sold domestically and overseas by a nation. The estimates of '$1.25 dollar a day' as reported by the United Nations in the MDGs above is an example of attempts to establish an international 'poverty' line and enable comparisons between countries. It is based on national assessments of the expenditure of households required to consume a certain number of calories or basket of goods and services. The 'purchasing power parity' adjustment widely attached to these assessments enables differences across countries in the costs of services and value of currencies and inflation rates to be accounted for.

In 2000, the World Bank undertook a 'participatory poverty assessment' with 40,000 poor women and men in 50 countries as part of the preparation of the World Development Report on *Attacking Poverty*. It was a significant departure from their previous work on poverty and involved working with focus groups, where people discussed how they themselves defined their sense of well-being and impoverishment. The multidimensional nature of 'poverty' and 'development' that the research revealed is evident in Figure 2.13, where factors of security, autonomy and self-esteem (as well as income) were important in shaping peoples' own definitions of well-being.

Other institutions and authors incorporate a number of components of 'human development' to consider poverty and deprivation. Since 1990, the United Nations Development Programme (UNDP) has reported a 'Human Development Index' (HDI) drawing heavily on the work of Amartya Sen (1981, 1999). Sen considers human development in terms of individuals' capabilities to achieve, to flourish and live lives they have reason to value. In turn, poverty is considered as a set of interrelated 'unfreedoms' that constrain people's choices and opportunity to exercise their individual agency. The HDI (Figure 2.14) combines four indicators that measure three dimensions considered essential to enlarging people's choices and opportunities to engage in processes of human development: the

Figure 2.13 *Well-being as revealed through participatory poverty assessments*

● **Material well-being:** food, shelter, clothing, housing and certainty of livelihood in terms of possession of assets

● **Physical well-being:** physical health and strength (in recognition of how quickly illness can lead to destitution)

● **Security:** peace of mind and confidence regarding personal and family survival. Includes issues relating to livelihood, but also war, corruption, violence, lack of protection from police and access to justice and lawfulness

● **Freedom of choice and action:** power to control one's life, to avoid exploitation, rudeness and humiliation. Ability to acquire education, skills, loans and resources to live in a good place. Having means to help others and fulfil moral obligations

● **Social well-being:** good relations with family and community, being able to care for elderly, raise children, marry. Ability to participate fully in community/society, in gift-exchange, festivities, weddings, etc.

Source: Narayan et al. (2000).

ability to lead a long and healthy life, to be educated, and to enjoy a decent standard of living. In 2010, two new measures were introduced (see Box 2.5) in recognition that inequality and exclusion are also important aspects of poverty. To ensure human development is sustained over time and across generations, the UNDP considers it essential to address current equity and empowerment concerns. Understanding poverty and the chance to lead a meaningful life demands consideration of the distribution of advantage in society and possibilities to participate in decision-making.

Different definitions and conceptions of poverty are important, influencing who is identified as poor and the policy measures undertaken to support poverty alleviation, for example. As summarised by White (2008: 25):

> when poverty is defined solely in terms of income, then it is unsurprising that economic growth is found to be the most effective way to reduce poverty. But if basic needs such as health and education are valued, then the development strategy is likely to put more emphasis on social policy.

Each definition of poverty has its limitations. Monetary indicators such as GDP per capita are relatively easy to measure and data is accessible. However, its failing in terms of capturing the 'human' dimensions of development or the non-income dimensions of poverty have been long-debated in development studies. Recently,

Figure 2.14 *The components of the Human Development Index*

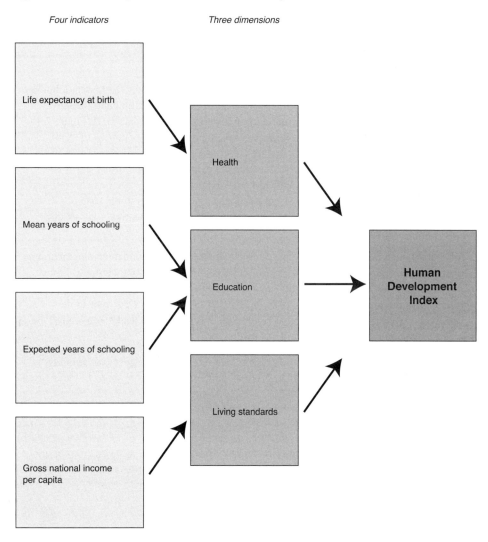

Source: adapted from UNDP (2010).

concern has also been expressed within 'developed' countries over
the adequacy of GDP as the conventional measure of 'progress'. In
the UK and France, for example, new measures are being developed
to monitor 'happiness' and 'prosperity' (see Chapter 6). There is
growing unease in particular over how GDP makes no distinction
between transactions that add to well-being and environmental health
or detract from it and enhance degradation.

Box 2.5

Measuring inequality and exclusion in human development

The Inequality-adjusted Human Development Index (IHDI) captures the losses in human development due to inequalities in the distribution of health, education and income. Across all countries, the IHDI was found to be 22 per cent lower than HDI. Figure 2.15 confirms that inequality compromises human development across countries of very contrasting levels of HDI and that there are large differences in inequality within these HDI groups. Evidently, higher levels of human development do not ensure parity in opportunities (as seen in South Korea and Peru) and greater equity can be achieved in the context of overall quite low levels of human development, as seen in the case of Ghana. In more than a third of countries, inequalities in health and/or education were greater than those of income.

The Gender Inequality Index (GII) is a measure of the losses in human development due to the disadvantages facing women and girls that combine to disempower these groups. The measure combines five indicators, as seen in Figure 2.16. Losses in human development due to gender inequality were found to range from 17 per cent in the Netherlands to 85 per cent in Yemen. Differences in reproductive health constituted the largest contributor to gender inequality globally.

Figure 2.15 *Loss in HDI due to multidimensional inequality (largest and smallest losses across HDI groupings)*

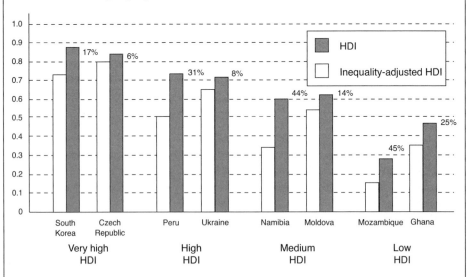

Note: Numbers beside bars are percentage loss due to multidimensional inequality.

Source: adapted from UNDP (2010).

Figure 2.16 *Components of the Gender Inequality Index (GII)*

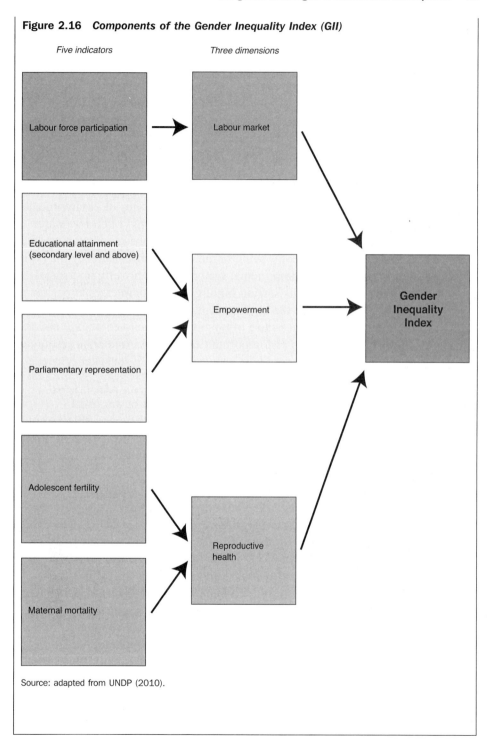

Source: adapted from UNDP (2010).

Poverty and the environment

Poor people have been regularly portrayed as both 'victims' and 'unwilling agents' of environmental degradation. In short, living in poverty can restrict the options people have for resource management: they may have to cultivate marginal lands, live in unsafe housing or remove remaining woodlands in order to sustain their household in the short term, with possibly detrimental effects on the resource base and their own longer-term livelihoods. Their environmental concerns are often those associated with immediate survival needs, such as for fuel, access to clean water and sanitation, or in securing productive lands. As such, processes of environmental decline (and the impacts of sudden-onset hazards) often hit poorer groups hardest. As discussed in Chapter 1, the impacts of environmental degradation are currently being borne disproportionately by the poor and at times this is the principal factor causing poverty and social conflict. However, this section highlights the complex relationship between poverty and the environment. This includes understanding the multi-dimensional nature of poverty (as discussed above) and the diversity of functions and opportunities that environmental resources and ecosystems can provide in securing not only subsistence needs, but wider well-being and routes out of poverty.

Worldwide, across rural and urban areas, the poor are often concentrated in environments that are 'inherently' poor, that is, are ecologically marginal and require high levels of investment in order to remain productive or be safe for human habitation (the resources that low income groups often don't have). In the mid-1980s, research suggested that 57 per cent of the rural poor and 76 per cent of the urban poor were resident in areas where ecological destruction and/or severe environmental hazards threatened their well-being (Leonard, 1989). In 2000, an estimated 1.3 billion people lived in what were termed 'fragile' lands (World Bank, 2003a) defined on the basis of soil characteristics, aridity and slope constraints (or combinations thereof). Forty per cent of the population of sub-Saharan Africa and one quarter of all people in Asia for example were resident in such fragile areas. Furthermore, the research found that these areas were home to an increasing proportion of the total population of developing countries (doubling since 1950) and a majority share of people in extreme poverty, suggesting a 'geographical retreat' of poverty into more impoverished and ecologically marginal areas over time. Similarly in urban areas, low income groups are often increasingly concentrated in inherently poor locations such as on

Plate 2.1 *The challenges of aridity to human settlements*

a) Northern Nigeria

Source: Hamish Main, Staffordshire University.

b) Southern Tunisia

Source: author.

steep hillsides and river beds prone to flooding and mass movements. In addition, they regularly reside in environments impoverished through their 'acquired' characteristics, such as alongside hazardous installations or railway tracks. This often reflects the operation of market forces, these environments having low commercial value (i.e. demand from other uses) because they are poor. For low-income groups who are often also politically marginalised, this may enable them to afford to live there and be able to avoid eviction.

Poorer groups in society also have less power to resist and prevent 'detrimental developments' – those that make their environments more impoverished in some way. As identified in Chapter 1, notions of environmental justice developed in the US context that centred on issues of race and class, have more recently been extended to include concerns for the capacity of low-income groups worldwide to influence decisions that shape the distribution of and access to both environmental 'goods' and 'bads'. The climate justice movement considered below raises a central question of how developing countries as a whole contribute very little to greenhouse gas emissions yet are already suffering from and are likely to be impacted most by further climate change. Box 2.1 illustrated how toxic waste materials continue to be transferred from higher incomes countries to generally lower income countries and how it is regularly the poorest who suffer environmental damage and health impacts through such processes. People in the most disadvantaged areas across England (identified using the Index of Multiple Deprivation) have also been shown to suffer the least favourable environmental conditions including proximity to waste management facilities, industrial and landfill sites, greater flood risk and poorer air quality (Walker et al., 2005). The lack of voice that poor people often have in environmental decision-making has been confirmed in China: for citizens with similar levels of pollution exposure and education, those living in high-income provinces were more than twice as likely to file complaints as those residing in low-income provinces (WRI, 2003). There are many further examples throughout this text of 'development' processes undermining both environmental and human health, and empowering local voices to enable communities to resist detrimental developments is seen in Chapters 4 and 5 to be a critical feature of more sustainable development processes.

Significant steps in understanding the complex linkages between poverty and the environment have been made recently, including through the work of the Millennium Ecosystem Assessment (MEA,

2005). As seen in Chapter 1, the MEA was a very detailed and extensive consideration of the current patterns and drivers of ecosystem change across all the principal biomes of the world. The conceptual framework used in the MEA is shown in Figure 2.17. Human well-being was considered to have five major components: security, basic material needs, health, good social relations and freedom of choice and action. Freedom of choice and action is identified as being both influenced by and a pre-condition for the other components of well-being. Ecosystems are seen to underpin human well-being through a number of 'services' that they provide, in regulating, provisioning, supporting and cultural services. Clearly, human well-being also depends on other factors such as quality of education and health systems, aspects of technology and the activities of institutions and quality of governance. Some aspects of ecosystem services can also be influenced or 'mediated' by socio-economic factors (such as the possibility of purchasing substitutes for particular services or engineering flood defences). As a result, the strength of the relationships between ecosystem services and human well-being will be different in different ecosystems and according to socio-economic circumstances. The arrows within the figure depict the broad intensity of linkages and potential for human intervention.

The assessment confirmed that 'ecosystem services are a dominant influence on the livelihoods of most poor people' (MEA, 2005: 61). Approximately 70 per cent of the world's poor live in rural areas and are highly dependent on the 'provisioning' services of ecosystems; in supporting food production (including in agriculture, livestock, fishing and hunting) and for primary sources of energy (e.g. woodfuel) and for protecting and providing water resources. The MEA also confirmed that poor people are highly vulnerable to changes in ecosystems, most obviously when such provisioning services are threatened (such as through enhanced water scarcity), but also through impacting on other components of well-being. Freedom of choice and action, for example, is compromised through ecosystem decline when people (often women and children) have to spend more time in the collection of water and fuelwood for basic household needs, thereby reducing time for education, care of family members and employment. In societies where cultural identities may be closely linked to particular local environments, habitats and wildlife (such as in the case of pastoral nomadic societies and several Arctic populations), the loss of 'cultural' services through ecosystem decline impacts very starkly on well-being (and cannot be readily substituted or replaced).

Figure 2.17 Linkages between ecosystem services and human well-being

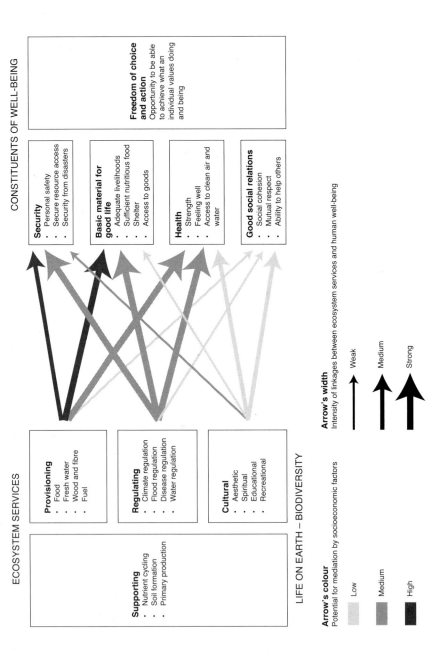

ECOSYSTEM SERVICES

CONSTITUENTS OF WELL-BEING

Supporting
· Nutrient cycling
· Soil formation
· Primary production

Provisioning
· Food
· Fresh water
· Wood and fibre
· Fuel

Regulating
· Climate regulation
· Flood regulation
· Disease regulation
· Water regulation

Cultural
· Aesthetic
· Spiritual
· Educational
· Recreational

LIFE ON EARTH – BIODIVERSITY

Security
· Personal safety
· Secure resource access
· Security from disasters

Basic material for good life
· Adequate livelihoods
· Sufficient nutritious food
· Shelter
· Access to goods

Health
· Strength
· Feeling well
· Access to clean air and water

Good social relations
· Social cohesion
· Mutual respect
· Ability to help others

Freedom of choice and action
Opportunity to be able to achieve what an individual values doing and being

Arrow's colour
Potential for mediation by socioeconomic factors

Low
Medium
High

Arrow's width
Intensity of linkages between ecosystem services and human well-being

Weak
Medium
Strong

Source: MEA (2005).

The MEA also found that with diminished human well-being, dependence on ecosystem services increases, creating additional pressures on ecosystems and potential for further degradation of ecosystem services. Moreover, many of the regions and groups facing the greatest problems related to the sustainable supply of ecosystem services were also found to be those facing the greatest challenge in achieving the Millennium Development Goals, confirming the close interdependence of challenges of environmental conservation and poverty alleviation.

Work in the field of political ecology has also done much to explore why the poor are so often marginalised ecologically, economically and politically. Political ecology explores the deeper underlying factors such as those of political structures and economic and social relations that in combination with the characteristics of ecology shape individual resource management decisions. This work has put issues of power and control over resources central to understanding many issues of resource degradation, vulnerability to hazards and environmental change (see Blaikie and Brookfield, 1987; Peet and Watts, 2004), as well as to understandings of poverty. In all these arenas it has become important to understand the varied 'command' over environmental resources which people have. This can be through their individual ownership or via membership of particular social groups, but also according to such things as their gender. Different groups of people also have varied capacities to withstand environmental stresses and shocks and ability to make sustainable use of these different 'environmental entitlements' (Leach and Mearns, 1991). Figure 2.18 illustrates how questions of power and control in resource use and environmental change extend to within households. This work is now very important for understanding the vulnerabilities of people to move into and out of poverty over time (central to the achievement of the MDGs) as well as their resilience to environmental changes (the capacity to absorb shocks and stresses and to recover but also to take new opportunities for development). Both these concepts of vulnerability and resilience are investigated further in relation to adaptation to climate change below.

Evidently, there are significant global challenges in understanding and responding to the multidimensional (and multilevel) problems of poverty for sustainable development. Whilst the environmental concerns and development needs of the poor are certainly often in stark contrast to those of more 'wealthy' groups or 'northern' priorities, and typically are associated with securing the most basic

Figure 2.18 *The gendered experience of environmental change*

Saltwater intrusions/salinity of groundwater is a major problem for people living in southwestern districts of Bangladesh. Women are generally responsible for collecting water and often travel 5–10 km each day on foot. This has implications for women's time and makes it difficult to fulfil other household roles. It also creates tension within households, with cases of physical abuse from husbands. Women suffer more illnesses through the strain of travel, especially when they are pregnant and shortly after giving birth. Environmental change is also transforming local customs. For example, aged parents are fearful of daughters getting married and leaving them with no-one to fetch drinking water. Some people living in areas where the salinity crisis is more moderate are refusing to arrange marriages for their offspring with people from worse-hit areas.

Research on the impacts of cyclones and resultant flooding in Bangladesh in 1991 found that the death rate for women was five times higher than for men. This was considered to be linked to gender roles and cultural norms that restrict women's mobility. Men were able to warn each other as they met in public spaces but often did not communicate information to the rest of the family. Many women are not allowed to leave their home without a male relative and when floods hit, they waited for relatives to return home and take them to safety. As in many Asian countries, most Bengali women have never learned to swim.

Source: compiled from IIED (2007).

levels of economic and social well-being, the linkages between poverty and the environment are complex and rarely direct, ensuring that there are no simple blueprints for sustainable development, as will be illustrated in Chapters 4 and 5. However, this section confirms that poverty clearly denies millions of people basic rights in the short term as well as their potential to achieve their development aspirations in the future.

The environment cannot cope

Perhaps the starkest realisation of the need to find new patterns and processes of development has come from an understanding of the environmental unsustainability of contemporary development. At the time of the Stockholm conference in 1972, the primary environmental problems tended to be national: 'the environmental sins of one nation did not generally impinge upon other nations, let alone upon the community of nations' (Myers and Myers, 1982: 195). Ten years later, cross-border environmental problems which affected many nations, such as acid rain or nuclear waste, were being recognised. By the time of the Earth Summit in 1992, a series of environmental problems that affected the global community as a whole were identified in two senses. First, those human impacts on the environment that have a 'supranational' character such as climate change and ozone depletion through the connectedness of atmospheric systems where change generated anywhere has potential effects around the globe. Second, local issues such as the loss of forest, soil erosion or accessing clean water were occurring repeatedly in numerous locations (on a worldwide scale), thereby posing threats to resources on which more and more people of the globe depend.

The ongoing global challenge of addressing many local-scale environmental issues reproduced worldwide was confirmed within the MEA. Whilst substantial expansion in 'provisioning services' has been achieved over the last 50 years, particularly in crop, livestock and fisheries production, these have been achieved principally at the cost of the degradation of other ecosystem services. Problems of over-extraction of ground water and soil degradation are identified as persistent local problems worldwide. However, it is climate change that is understood currently to present the most complex challenges for sustainable development that 'span science, economics and international relations' (UNDP, 2007: 4). Climate change is the 'archetypal' global challenge where a 'a molecule of greenhouse gases emitted anywhere becomes everyone's business' (Clayton, 1995: 110). However, there is substantial uncertainty and contestation concerning how the climate is changing, about how human society (and natural systems) will respond and about the resultant impacts for environment and human development.

In previous decades, climate change was considered principally a concern for high-income countries and early actions focused on the 'mitigation' or moderation of the impact for future generations. Such actions include the targets for the reduction of greenhouse gas emissions encompassed in the Kyoto Protocol, as identified in Box 2.6. It is now understood that climate change is impacting on many ecosystem services and local level environment and development opportunities within current generations and particularly in low-income countries. In short, the climate change and development agendas have moved closer together in recent years. This includes understanding that the emerging risks of climate change will 'fall disproportionately on countries that are already characterized by high levels of poverty and vulnerability' (UNDP, 2007: 25) but also that climate change questions 'development' itself worldwide. As seen in the sections above, moving towards lower-carbon economic growth is a contemporary concern internationally. There is also close consideration worldwide as to how addressing the

---

Box 2.6

The development of an international framework on climate change

1992: The United Nations Framework Convention on Climate Change (UNFCCC) – the 'Climate Convention' – signed by 167 countries at the Rio Earth Summit, came into force in 1994. A 'soft' law that called on parties to the convention to voluntarily commit to emissions reductions towards the stabilisation of emissions at 1990 levels by 2000.

1997: One hundred and thirteen signatories to UNFCCC agree to the Kyoto Protocol. This established the principal of legally binding targets for greenhouse gas emissions reduction for 'Annex 1' countries (the OECD members plus countries of the former USSR and Eastern Europe). The Kyoto Protocol commits developing countries to monitor further and address their emissions towards climate change mitigation. The Protocol also established the principle that some emissions reduction requirements could be traded and transferred between countries through a number of 'Flexibility Mechanisms', thereby introducing an official market in carbon.

2001: The Marrakech meeting of the Conference of Parties (COP) formally recognises the challenges of adaptation to climate change for developing countries and links the participation of developing countries in future climate agreements to finance and technology transfers from richer nations to support their adaptation needs.

2005: The Kyoto Protocol becomes fully ratified and legally binding. Thirty-seven Annex 1 countries are legally committed to reducing their emissions by an average of 5% below 1990 levels by 2012. The US and Australia are not signatories nor subject to these targets. It expires in 2012.

2009: The UN Climate Conference 'COP15' meeting in Copenhagen is held to consider the post-Kyoto framework for international commitments on actions to address climate change beyond 2012. Key areas of debate at the conference included:

● The overall goal for limiting the increase in global temperature and therefore the level of action required (representatives from the Small Island States for example, argued for 1.5 degree increase rather than the 2 degrees favoured by the G20).
● Which countries in future should be subject to binding emissions targets (whilst Kyoto placed different responsibilities on 'developed' and 'developing' countries, a broader consideration of emissions targets is understood to be needed, particularly as emissions from BRIC countries rise).
● How to account for efforts to reduce deforestation and degradation (i.e. that support carbon sequestration) within targets.
● Future funding of adaptation efforts in vulnerable countries.

There was no legally binding outcome from Copenhagen. Rather, 25 parties to UNFCCC agreed to the Copenhagen Accord, the remaining parties noting the 'existence of the accord'. Agreement was made to hold increase in global temperatures below 2 degrees

Celsius but no new emissions targets were negotiated (rather parties will 'pledge' voluntary, 'nationally appropriate' targets that will be internationally reviewed and monitored). The Accord proposed short-term funding of US$30 billion to be made available between 2010 and 2013 to support adaptation activities. A High Level Panel was also established to identify sources and mechanisms to deliver long-term finances from developed to developing countries for climate adaptation. This includes consideration of payments for the protection of forests.

Figure 2.19 *The merging of climate and development agendas*

- African Development Bank (2003) *Poverty and Climate Change: Reducing the Vulnerability of the Poor through Adaptation*
- UN Development Programme (2007) *Fighting Climate Change: Human Solidarity in a Changing World*, Human Development Report 2007/8
- UN Environment Programme (2007) *Environment for Development*, Global Environment Outlook 4
- Department for International Development (2010) *The Future Climate for Development: Scenarios for Low-income Countries in a Climate-changing World*
- World Bank (2010) *Development and Climate Change*, World Development Report

climate challenge could also deliver opportunities for more sustainable development as the issues of responding to the global economic crisis become apparent. Figure 2.19 highlights a number of recent reports of major institutions that have explored these common agendas.

Evidently, the challenges that climate change presents for sustainable development are extensive (as well as often uncertain and heavily contested). Climate change is not a discrete environmental issue (as in the case of acid rain for example), but embraces interlinked concerns for environmental integrity, economic development and social justice now and into the future. The following sections identify a number of key features of the challenges of climate change for sustainable development. In Chapter 3, the way in which these are being addressed in action are considered, with the outcomes for particular people and environments following in Chapters 4 and 5.

Climate warming confirmed?

The Earth's climate system is extremely complex but there is little doubt that it is continuously under change. What is widely debated is whether it is warming (and if so, how quickly), what the driving factors of change are and what could be the projected impacts. In 1988, the UN and the World Meteorological Office established the Inter-governmental Panel on Climate Change (IPCC) to provide a scientific view on these questions. The IPCC reviews and assesses the research produced worldwide on climate change with thousands of scientists contributing to their work. In 2001, the IPCC established that the climate was warming: global mean surface temperatures had increased by 0.6 degrees Celsius (+/–0.2 degrees) since the late nineteenth century. It also reported that most of the observed warming was due to human activities (as seen in Figure 2.1). By 2009, the evidence for warming was considered 'unequivocal' and the mechanism underpinning climate warming was established: human activities were leading to rising concentrations of greenhouse gases in the atmosphere and driving the identified 'enhanced greenhouse effect'.

'Greenhouse gases', including water vapour, methane, nitrous oxide, ozone and carbon dioxide, regulate the temperature of the Earth by controlling the re-radiation of solar radiation back into space; serving to keep the Earth around 33 degrees Celsius warmer than it would be. The IPCC established (through direct measurements over the last 50 years plus measurements of historic levels within ice cores) that concentrations of carbon dioxide moved very starkly from a long period of stability to exponential growth from around 1750. These changes in the chemical composition of the atmosphere are causing the Earth to trap more heat in the atmosphere and therefore to warm up. Carbon dioxide is the principal greenhouse gas in that it provides the main warming or 'forcing effect' in the atmosphere. At a global scale, the major source of carbon dioxide is the burning of fossil fuels, contributing approximately 75 per cent of human-caused carbon emissions since the 1980s (IPCC, 2001). Once emitted, CO_2 and other greenhouse gases stay in the atmosphere for a long time, ensuring that even if emissions were to stop today, warming will continue.

Whilst there is relatively little dispute over the rising greenhouse gas concentrations and the effect on global mean surface temperatures, it is less confidently understood how such increases in surface temperature are affecting (and will go on to affect) the various complex components of the climate system to impact on weather patterns, ocean currents and the biosphere for example. Figure 2.20 identifies some of the sources of uncertainty in knowledge regarding the complexities of climate change. These uncertainties challenge climate scientists but are also the focus for climate change 'sceptics'

Figure 2.20 *Ongoing questions for climate science*

- What are the relative roles of natural and anthropogenic radiative forcing mechanisms in climate change (little is known for example about changes in solar activity or impacts of land use change on surface albedos)?
- What interactions exist between different elements of the climate system (for example, the impact of atmospheric changes on oceanic circulation)?
- What are the effects of feedback loops: 'positive' feedbacks, for example, with melting of ice and snow, less solar radiation is reflected, thereby leading to greater warming; and 'negative' loops that include higher carbon dioxide concentrations speeding up plant growth and carbon uptake, thereby reducing CO_2 concentrations?
- What is the potential of 'tipping points' in the climate system being reached in future (where a critical threshold in the control of the system is reached, leading to total loss and breakdown)? Potential elements include the loss of the Greenland ice sheet and shutdown of the gulf stream in the North Atlantic.
- What are the different time-scales at which these processes occur?

Source: compiled from Peake and Smith (2009).

or even 'deniers' (a controversy that is accepted to be more evident in the popular media than it is within the scientific communities and literatures).

Such 'scientific' gaps also combine with substantial uncertainties regarding the future evolution of societies: 'the effects of human activities on climate depend on future emissions of greenhouse gases, and the impacts of the resulting changes in climate depend on the future state of the world' (Arnell et al., 2004: 3). The IPCC has modelled six different 'emissions scenarios' based on different projections of how society, economy, population and political structures may change in future decades. These different 'development pathways for the world' (ibid.) included projections of land use change, economic growth rates, technological innovation and attitudes to social and environmental sustainability, for example. Significantly, they do not take account of any deliberate action to reduce emissions (such as the impact of the Kyoto Protocol). Each emission scenario is modelled to have different outcomes in terms of further temperature change and in turn will have varied consequences for environmental systems and human societies. Figure 2.21 summarises the impacts on various systems, sectors and regions relative to the projected temperature increases based on these scenarios. It confirms that even small changes in surface temperatures challenge societies worldwide to cope with and adapt to changes in, for example, food systems, water availability and marine ecosystems. With larger increases, the risks and burdens become more extensive and serious. The IPCC identifies 2 degrees Celsius of warming as the point at which the risk of large-scale human development setbacks and ecological catastrophe increase sharply. To avoid such 'dangerous anthropogenic climate change', the IPCC suggests that global emissions need to be reduced by at least 80 per cent over 1990 levels by 2050. This figure of a maximum 2 degrees increase in global average surface temperatures is the target used in current international climate negotiations, as seen in Box 2.6.

Figure 2.21 *Summary of projected impacts of increases in global mean surface temperature*

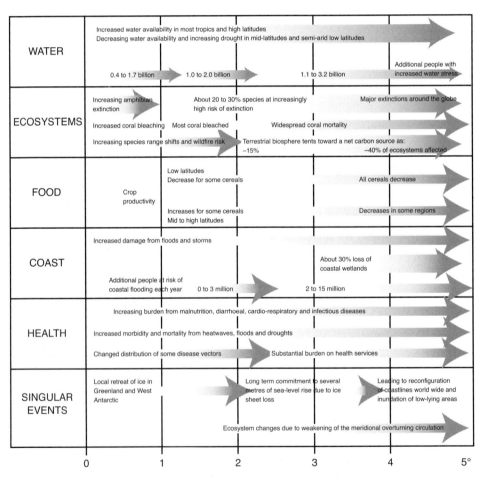

Global mean annual temperature change relative to 1980–1999 (°C)

Source: adapted from Houghton (2009), based on IPCC (2007).

Importantly, the IPCC (2001) identified that the impacts of climate change would not be manifested in the same way in all regions. Small island states and coastal and delta regions will be especially affected by the increased risks of coastal and river flooding and storm surges, as identified in Figure 2.21. Increased water stress and drought are predicted to impact most profoundly in mid latitudes and semi-arid low latitudes. Furthermore, the report confirmed that developing countries are likely to suffer most from the negative impacts of climate change. This is due to their geographical location in regions with existing high levels of climate variability, but also because of the economic importance of climate-sensitive sectors (such as agriculture and fisheries) within their economies and individual livelihoods. These countries also generally have more limited human, institutional and financial capacity to anticipate and respond to the impacts of climate change. The report confirmed that mitigation efforts will not prevent climate change from occurring and closer attention is needed to adaptation efforts. These key terms are defined in Figure 2.22. The IPCC urged particular attention be given to adaptation in developing countries where climate change brings additional threats and compounds existing risks and vulnerabilities within countries with the fewest resources and where the majority of the world's poor live. In short, the 2001 IPCC report was very important in bringing the development agenda into the climate change arena. It has helped shift attention to climate change adaptation and efforts to support the development of adaptive capacity within low-income countries that are explored further in the climate justice section below.

Figure 2.22 *Key terms in responding to climate change*

Mitigation: an anthropogenic intervention to reduce the sources or enhance the sinks of greenhouse gases.

Adaptation: adjustments in natural or human systems in response to actual or expected climatic stimuli or their effects, which moderates harm or exploits beneficial opportunities.

Adaptive capacity: the potential and capability to change to a more desirable state in the face of the impacts or risks of climate change.

Source: compiled from IPCC (2001).

The political and economic challenges

Whilst the scientific evidence for human-induced global warming has mounted, it is suggested that political responses have been slower to emerge (UNDP, 2007; Giddens, 2009). As climate change is not a single problem, political decisions across all sectors and on behalf of many different organisations are challenged to embed climate change considerations throughout their strategies and actions. How this is being developed in practice is considered in Chapter 3. However, Giddens (2009: 2) suggests a central paradox of climate change that ensures such political decision-making is difficult:

> since the dangers posed by global warming aren't tangible, immediate or visible in the course of day-to-day life, however awesome they appear, many will sit on their hands and do nothing of a concrete nature about them. Yet waiting until they become visible and acute before being stirred to serious action will, by definition, be too late.

Addressing climate change requires political leaders to take action now for what are often long-term outcomes (and where the benefits may fall principally in other geographical regions). Yet, future generations will already have to live with the consequences of current emissions due to the inertia within the climate system and 'with climate change, doing nothing offers a guaranteed route to a further build up of greenhouse gases and to a mutually assured destruction of human potential' (UNDP, 2007: 4). Furthermore, the challenges of climate change cannot be solved by one generation of political leaders as sustainable emissions trajectories will be needed for decades not years. The uncertainty of many aspects of climate, however, makes assessing the political risk of actions and strategies to tackle (unpredictable) problems hard to establish. Furthermore, action on climate change requires a strong role for governments in intervening in public life and business operations that is substantially different to previously dominant ideas about how development should occur, as seen in Chapter 1. A central challenge is for political leaders to convince their voters of the need for and basis of these policies and to maintain them when more immediate, local and visible issues arise. This was seen above in terms of maintaining public subsidies for clean energy in the context of the current economic crisis. At their core:

> Climate change policies require trade-offs between short-term actions and long-term benefits, between individual choices and global consequences. So climate change policy decisions are driven fundamentally by ethical choices. Indeed, such decisions are about concern for the welfare of others.
>
> (World Bank, 2010a: 52)

The ethical choices embraced by climate change are at the centre of the climate justice movement. Its work centres on issues of fairness and justice of the current impacts of climate change and as a result of the policies and strategies being undertaken to address climate change (considered further below).

The political challenges of climate change also include the need for political leaders to work internationally in often complex negotiations. As seen in Box 2.6, the Kyoto Protocol is an international agreement towards mitigating climate change through setting targets for emissions reduction. The Protocol recognised 'differentiated' responsibilities globally; only 'Annex 1 countries' (countries of the OECD and of the former USSR and Eastern Europe) were subject to legally binding emissions targets specific to their country situation, whilst the emerging economies were required to 'monitor further and address' their emissions. A key part of the Protocol was the creation of a mechanism for 'trading' in these 'permits to emit'. This was the foundation for a carbon market that is heavily contested as to its contribution to sustainable development (considered further below and in Chapter 3). Box 2.6 confirmed the difficulties of international political negotiations as seen in the most recent UN Climate Conference in Copenhagen in December 2009.

A further reason why political action has been slow to emerge, has been that much of the debate around climate change has tended to be in scientific arenas and in a 'language' not well understood by political leaders. In 2007, publication of the 'Stern Review' *The Economics of Climate Change* is considered to have changed this. Nicholas Stern (an economist formerly at the World Bank) was commissioned by the UK government to undertake an assessment of the costs of acting to avert climate change and those of 'inaction'. He concluded that if no action was taken, the global economy could shrink by 5 per cent each and every year (and up to 20 per cent), whereas the costs of acting to avert climate change were much smaller – an estimated 1 per cent GDP per annum to avoid the worst effects. He urged decisive near-term action to mitigate devastating future impacts. In particular he argued for a rapid expansion of the carbon market and an increase in the trading of emissions permits between nations. Stern's argument was based on the idea that as the costs of controlling greenhouse gas emissions are variable across the world, trading in permits enables 'comparative advantage' to be sought – investments are made where costs are lowest and thereby the overall economic costs of reducing emissions globally can be minimised.

Stern's work has been criticised in terms of both the economics and the science that it embraces. Stern himself in his latest book (2009) has suggested that his costings were conservative and that he underestimated the speed of climate change. However, Peake and Smith (2009: 191) suggest that the review 'achieved things that previous scientific and policy reports had not' by putting the climate challenge into the financial terms understood by prime ministers and presidents (not solely environment ministers) and taking the debate into new arenas such as meetings of the G20. For others, particularly in the development field, the Stern Review was short on strategies for action and underestimated the political difficulties in ascertaining responsibilities and the appropriate response.

Climate justice and the challenges of adaptation

Just as climate change has risen up the agenda for political leaders internationally in recent years, there has also been a wider politicisation of climate change emerging from many different 'grassroots' groups, organisations and coalitions that are increasingly working together transnationally under the broad banner of 'climate justice'. Their activities include public demonstrations (coordinated by organisations such as Plane Stupid! and Rising Tide), advocacy and training work (such as that promoted by the Environmental Justice and Climate Change Initiative based in the US) and looser networks of people engaging in periodic 'climate camps' that often receive wide media attention. Many of these groups came together within the Climate Justice Action Network formed in 2009 to draw global attention to the failings of the UN climate talks, suggested to be 'an unjust set of negotiations interested in expanding capitalism rather than in addressing the global climate crisis' (Building Bridges Collective, 2010: 27). For many groups, their concerns may not be explicitly 'climate' related, but based on varied local struggles against land dispossession, water rights or new urban developments, with activists North and South making links between their respective local concerns. Others challenge the dominance of powerful countries and corporations in international negotiations such as trade. Many have their origins in the wider anti-globalisation movement of the 1990s (considered in Chapter 1) where their shared concerns were in the deepening inequalities being caused by capitalism. In 2010, the Peoples' World Conference on Climate Change and Mother Earth's Rights was held in Cochabamba, Bolivia (substantially inspired by

President Evo Morales and in response to the failings of the COP15 meeting). The aim was to reaffirm the fight for climate justice, to expose the structural causes of climate change and to work for radical, alternative measures to ensure the well-being of all humanity in harmony with nature built on principles of solidarity, justice and Mother Earth's rights (http://pwccc.wordpress.com/). In short, the climate justice movement draws together those who connect the weak responses to climate change and the many injustices surrounding the problems of climate change to the dominance of capitalism. Box 2.7 expands on the key arguments underpinning climate justice that have helped put climate change centrally as a human rights issue and an issue of development itself.

Understanding climate change in terms of human rights and social justice has demanded closer attention to questions of climate change adaptation and processes towards 'climate resilient' development. Adaptation includes making adjustments for the impacts of climate change and supporting people's capacity to cope in future, as identified in Figure 2.22. To date, 'planned adaptations' have largely been undertaken in more wealthy nations. Typically these focus on building dams and physical structures that provide barriers against extreme events like storm surges and protecting housing and transport networks (including designing new) that can withstand the wider and slower onset issues such as higher rainfall and drainage associated with climate warming. These countries also have the resources to invest in assessing climate-related risks, in undertaking analyses of who is most vulnerable, for example, and in establishing early warning, training programmes and insurance systems that can reduce future vulnerability and protect those most at risk. These adaptations are often not available within developing countries.

However, as Adger et al. (2003: 181) suggest, it is important to understand that 'people of developing nations are not passive victims. Indeed, in the past they have had the greatest resilience to droughts, floods and other catastrophes.' As seen above, climate shocks (as well as other risks) already feature prominently in the lives of the poor and local communities often have a wealth of knowledge concerning environmental change and have been adapting to and coping with unpredictability in weather patterns and extreme events for many years (IIED, 2009). However, the concern is that 'climate change will steadily increase the exposure of poor and vulnerable households to

Box 2.7

The principles of climate justice

1. The rich world should take responsibility. Central injustices include:

- The poorest of the poor are being hit first and hardest through climate change. Climate change is considered to be causing an additional 150,000 deaths worldwide through flooding, malaria and malnutrition, the vast majority occurring in Africa and Asia.
- Impacts of climate change will and do interact with pre-existing social and economic vulnerabilities, that is, emerging risks fall disproportionately on countries already characterised by high levels of poverty and vulnerability that are least empowered to adapt to and change the processes underpinning climate change.
- It is over-consumption in rich countries (and wealthy southern elites) that causes climate change and is responsible for the principal share of current emissions (Figure 2.23).
- Financial support should flow from the North to the South for climate adaptation measures. This is based on the principle that the rich world owes the poor an 'ecological debt', in particular given its responsibility for the majority of historical emissions (Figure 2.24).

Figure 2.23 *Share of global CO_2 emissions, 2009*

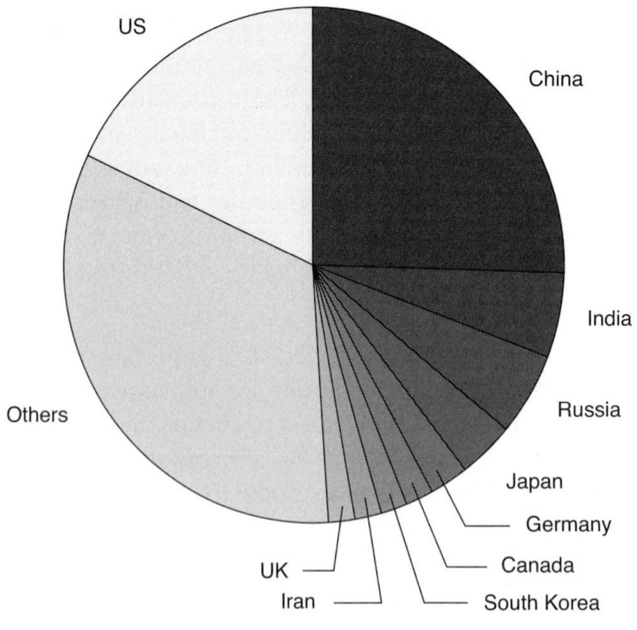

Source: Energy Information Administration (2009).

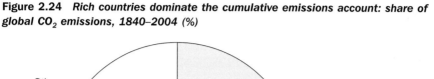

Figure 2.24 *Rich countries dominate the cumulative emissions account: share of global CO_2 emissions, 1840–2004 (%)*

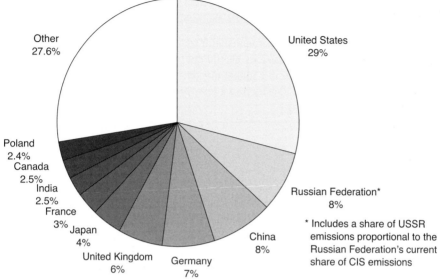

Source: adapted from UNDP (2007).

2. Climate change is caused by the burning of fossil fuels such that action on climate change must be focused on stopping it at this source.

   ● Climate justice campaigns for investments in energy efficiency and renewable energy technologies.
   ● Campaigns make linkages and remind people what drives the processes of industrialisation and over-consumption (by whom and who benefits) that are the root cause of climate change.
   ● The failures of climate policy are illustrated by the continued expansion of emissions per capita.

3. Challenging 'false solutions' to climate change.

   ● Climate justice movements campaign in particular around the aggressive promotion of carbon trading, biofuels (expanding into new geographical frontiers principally in the Global South) and geo-engineering solutions (such as carbon sequestration technologies) on the basis that these maintain the 'development as usual' approach and don't prevent further climate change or address poverty.
   ● Climate change results from our current and historical social relations and to address it, fundamental changes to our political and economic systems are needed.

- Carbon trading enables responsibility for actions to reduce emissions to be transferred from the North to the South with business and industry being the primary beneficiaries.

4. Natural resources should be conserved for the public good.

- Climate justice argues that environmental resources and resource functions such as the Earth's carbon cycling capacity should not be turned into property/commodities that can be bought and sold in a global market.
- The focus should be on a transition to a low carbon society that protects people's rights, jobs and well-being.

climate shocks and place increased pressure on coping strategies, which, over time, could steadily erode human capabilities' (UNDP, 2007: 17). More recent work on adaptation centres on understanding the underlying factors that cause vulnerability to climate change and on reducing these through development interventions that will enable adjustments not only to the risks associated with climate change but also other global environmental, social and economic changes, that is, through more sustainable development processes. The distinction between these two ideas of climate change adaptation is shown in Figure 2.25. Taking a vulnerability reduction approach ensures that adaption to climate change cannot be a 'stand alone' issue but must be integrated into core development strategies to deliver development processes that enhance resilience to all types of shock and that are also 'climate resilient'.

Figure 2.25 Different approaches to linking adaptation and development

Adaptation approach:

Adaptation to climate change impacts ⟶ Vulnerability reduction ⟶ Development

In this view, adaptation is carried out in response to the observed and experienced impacts of climate change on society (including ecosystems). These responses ensure that the vulnerability to the impacts is reduced. This in turn ensures that less is lost each time a climate-related hazard takes place, which means risk is reduced. With reduced risk, development can be more sustainable.

Vulnerability reduction approach

Development ⟶ Vulnerability reduction ⟶ Impact reduction ⟶ Adaptation

In this view, development processes help reduce vulnerability to climate change. By reducing the vulnerability, impacts of climate hazards are also reduced, as there is less sensitivity and exposure to the hazards. This translates into a process of adaptation to climate change.

Source: Schipper (2007).

Plate 2.2 *Challenging processes of development*

a) London, England

Source: Julie Doyle, University of Brighton.

b) Stuttgart, Germany

Source: Jolanta Gatzanis.

Conclusion

Almost 25 years since the WCED emphasised the urgency of the challenge of sustainable development, it is evident that environmental degradation continues to threaten human well-being, social cohesion and development opportunities worldwide and factors including population growth and rising resource consumption have placed increased pressure on the state of the global environment. However, this chapter has also highlighted how understanding of the interlinkages between these environment and development challenges has grown in recent years. The MEA, for example, has confirmed how environmental systems and services are the very basis for human well-being worldwide, but particularly for poorer groups and in regions where human development is currently lowest. Not only is poverty a challenge for the conservation of resources and environments, but it is now understood that environmental degradation threatens basic human rights and is a matter of justice when some people's opportunities for development are limited by the actions of others.

Perhaps most clearly, understanding of the challenges of climate change has opened a large space for reconsidering not only how human society is impacting on environmental systems of the globe, but the very meaning of human 'progress'. As the UNDP (2007: 15–16) suggests:

> There could be no clearer demonstration than climate that economic wealth creation is not the same thing as human progress . . . There could be no greater challenge to our assumptions about progress than that of realigning economic activities and consumption with ecological realities.

A key question is, if the current model of economic wealth creation is associated with mounting threats to human development currently and to the well-being of future generations, can this be considered 'progress'? As seen in this chapter, moving to lower carbon economic growth trajectories is understood to embrace challenges of change in production, transportation and consumption. Whilst new opportunities for economic development are also envisaged through technological innovations in energy efficiency and renewables, these also bring new environmental and political challenges.

Summary

Many of the core global challenges of sustainable development are long-standing; rooted in the resource-dependent nature of development, the persistence of poverty and rising inequality in terms of access to development opportunities and environmental improvements.

New challenges for sustainable development are presented by the rapid economic growth (and associated energy demands) in populous countries including China and India, but inequalities in 'how we live' currently can be argued to be more important in shaping the challenges of sustainable development.

There are multiple dimensions to 'poverty' and the relationship between poverty and the environment are complex.

Progress towards meeting the Millennium Development Goals is identified, but achievements are varied across geographical regions. Development processes are not meeting the immediate needs of the current generation, clearly illustrated by the persistent high levels of infant mortality.

Questions of power are central in understanding the persistence of environmental degradation and the prospects for sustainable development in the future.

The challenges of climate change raise complex questions for science, international relations, economists and development practitioners.

Discussion questions

- Discuss the various ways in which low-income groups may suffer more from environmental degradation than higher-income groups.

- What happens to your mobile phone when you update it for a fresh model?

- Is population growth the primary challenge for global sustainable development?

- Research the advantages and disadvantages of wind power as a renewable energy. How do these compare with those of biofuels?

- Discuss the strengths and weaknesses of the principles of climate justice.

- What do you think are the prospects of the current economic crisis to deliver new opportunities for sustainable development?

Further reading

Many of the texts in this Routledge Perspectives on Development series provide very accessible texts to follow some of the core challenges of sustainable development in more detail. They include:

Agnew, C. and Woodhouse, P. (2010) *Water Resources and Development*

Gould, W.T.S. (2009) *Population and Development*

Williams, A. and MacGinty, R. (2009) *Conflict and Development*

Young, E.M. (2011) *Food and Development*

For very clear reviews of the science of climate change and the challenges of mitigation and adaptation in practice see:

Houghton, J. (2009) *Climate Change*, Cambridge University Press, Cambridge.

Peake, S. and Smith, J. (2009) *Climate Change: From Science to Sustainability*, Oxford University Press, Oxford.

Both these texts have very good supporting illustrations and pointers for student investigations.

For recent data and reviews of research concerning issues central to the global challenges of sustainable development, see:

United Nations Development Programme (annual) *Human Development Report*, UNDP.

World Bank (annual) *World Development Report*, World Bank, Washington.

Worldwatch Institute (annual) *State of the World*, Earthscan, London.

Websites

www.climate-justice-action.org   A transnational network for people and groups committed to promoting and strengthening the rights and voices of affected people in confronting climate change. Site for exchange of ideas and resources.

www.ipcc.ch   Website for the Intergovernmental Panel on Climate Change – the international body that reviews the science of climate change. The IPCC does not undertake research or monitor climate change directly, but this is a portal for very influential publications, documents and details of current activities.

www.steps-centre.org   A well-established interdisciplinary research and policy centre that is working to bring development studies and science and technology closer together to address challenges of poverty and sustainability.

www.maweb.org/   Home page for the Millennium Ecosystem Assessment and the extensive research and reports regarding the current condition and trends of the world's ecosystems, the services they provide and the consequences for human well-being.

3 Actors and actions in sustainable development

Learning outcomes

At the end of this chapter you should be able to:

- Identify a number of institutions and organisations and the actions they are taking towards more sustainable development
- Be familiar with a number of key mechanisms for delivering environmental policy
- Understand how different organisations have changed the way they work (and with whom) as understanding of the demands of sustainable development in practice have evolved
- Understand the significance of international negotiations on trade, aid, debt and climate change in fostering more sustainable development and the factors that make these efforts difficult

Key concepts

Governance; multilateral environmental agreements, foreign aid; structural adjustment; environmental mainstreaming; foreign direct investment; corporate social responsibility; carbon offsetting

Introduction

It is evident through the previous chapters that taking action towards new patterns and processes of development that are more sustainable is a challenge for all levels of society, from individuals to the global community. This chapter identifies some of the actors, institutions and organisations considered key to shaping future sustainability: the international environmental and financial institutions, governments, business and civil society. Concurrently, a number of core arenas that explicitly link the prospects for sustainable development in less developed countries with actors and activities beyond their boundaries are also investigated: those of aid, trade and debt. The chapter considers the broad kinds of activity that they are undertaking and identifies how these have changed as understanding of the demands of sustainable development in practice have evolved. Of particular importance is the way that these different actors in practice increasingly work together, across different scales and through varied mechanisms in the pursuit of sustainable development.

From government to governance

Whilst historically, national governments took the predominant role in decisions regarding environmental management and human development concerns, it was seen in Chapter 1 that this position has been substantially challenged, in particular by the emergence of international environmental problems and processes of economic globalisation. Currently the actions that governments take are significantly shaped by international agreements (such as the Kyoto Protocol) and international financial institutions (such as the IMF), the decisions of private business investors and corporations as well as the activities of NGOs and citizen groups. Importantly, all these institutions now shape the processes through which change occurs, that is, they influence how decisions are developed and the mechanisms for action to shape the outcomes for sustainable development.

The term 'governance' is increasingly used across many arenas to recognise the range of 'institutions, rules and participants' that now operate in complex networks and across scales to 'steer societal concerns' (Peake and Smith, 2009: 239). Whilst it is a difficult concept, it is thought to have particular relevance in relation to sustainable development (see Lemos and Agrawal, 2006; Adger and Jordan, 2009; Evans, 2012). As seen, many environmental problems are affected by decisions, interventions and behaviours at local, regional, national and transnational levels and have impacts across those spatial scales. Furthermore, the driving factors behind environmental problems are many, are interconnected in complex and uncertain ways and the causes are often de-coupled in both time and space from their impacts (seen particularly clearly in relation to climate change). The concept of environmental governance captures this 'multiscalar' character of environmental problems. It also embraces the many kinds of institution (both 'formal' organisations such as the state, NGOs or corporations and the many 'informal' arrangements that people agree to such as within families and as part of civil society) that are understood to influence environmental actions and outcomes. The mechanisms and 'rules' of environmental governance through which change is sought now include international and national legislation, but also public sector taxes and subsidies that can restrict or incentivise particular kinds of environmental behaviour as well as the voluntary undertakings of individuals.

In previous chapters it was seen that particular actors and mechanisms have been prioritised towards achieving environment and development goals at particular times (and within particular political-economic and environmental contexts). It was seen in Chapter 1 that the 'first generation environmental problems' such as air and water pollution within countries were addressed principally by national governments through 'top-down' legislation and taxation to regulate business and industry and to limit particular kinds of polluting activity, for example. This so-called 'command and control' approach to environmental policy also extended to governments participating in Multilateral Environmental Agreements (MEAs) to address emerging international environmental issues such as acid rain. Through the 1990s the role of private sector organisations and market mechanisms to deliver environmental improvements became more widely used (as ideas of ecological modernisation dominated mainstream thinking on the environment). These were designed to adjust the economic incentives to business and individuals to encourage more sustainable production and behaviours. Over the last 20 years, understanding has risen of the need for and potential of a much broader array of what are termed 'hybrid governance strategies' (Lemos and Agrawal, 2006). These recognise the strengths and weaknesses of particular organisations and mechanisms to promote sustainable development in isolation and therefore emphasise partnerships and co-management (such as between the state and community) and a combination of what can be achieved through legislation, market incentives and self-regulation. Figure 3.1 identifies a number of such hybrid strategies currently used to foster more sustainable development at a range of scales. Many of these approaches are discussed in further detail through this and subsequent chapters.

Notions of environmental governance are also important in understanding sustainable development in that they help put focus on the processes of decision-making and the trade-offs and compromises

Figure 3.1 *Forms of environmental governance and policy instrument*

Forms of governance	Conceptual basis	Market component	Role of public authorities	Examples	Issues
Command and control	Coercive regulation of behaviour of business and individuals through legislation specifying standards of pollution that a process or product has to meet	None: requires each firm and person to implement same measure irrespective of relative costs to them to undertake measures	State (and international) organisations set, monitor and enforce standards backed up by legal system	Remains most widely used mechanism Emissions standards limiting gas releases from factories, vehicle exhausts, agricultural discharges to water courses Legislation to ban/phase-out products or date by which pollution control device in place Clean Air Acts banning use of coal, Montreal Protocol to phase out ozone-depleting substances, catalytic converters required in cars	Regulations costly to administer and enforce Blocks innovation as no incentive to reduce pollution below that required by law Economically inefficient as some polluters can reduce pollution more cheaply than others
Cap and trade	Ceiling set on total amount of pollutant allowed	Changes are made where and by whom they are most cost efficient to do so	International agreements and national legislation set overall limits for pollution in a certain time period	Used widely in national energy industries (trade in SO_2, Pb and NOx)	Scientific and political challenge of setting sufficiently ambitious 'cap'

continued …

Figure 3.1 ... continued

Forms of governance	Conceptual basis	Market component	Role of public authorities	Examples	Issues
	Quotas and permits to pollute are tradable	Market forces of supply and demand shape trading	Public authorities monitor and enforce	Newer markets in carbon – specifically Clean Development Mechanism of Kyoto Protocol and EU Emissions Trading Scheme	Volatility of markets hampers investment decisions Trading doesn't require emissions reductions and doesn't address root cause of environmental problems Questions of long-term outcomes for environment and social justice
Environmental taxes/pollution charges	Based on Polluter Pays Principle that cost of preventing pollution or minimising environmental damage should be borne by those responsible for pollution Holds that price of goods should fully reflect total cost of production including use of public goods (land, water, air) for emissions that are currently underpriced in existing markets	Environmentally damaging behaviours become more costly and therefore are avoided by firms and individuals	Government and public authorities intervene in market, set and collect charges/taxes and make decisions regarding allocation of revenues from taxes	Host of examples including: – Taxes on discharge of pollutants e.g. use of leaded petrol, airline flights – User charges e.g. for vehicle use in cities and fresh water metering – Taxes/charges on use of harmful products e.g. on pesticide sales and plastic bags in shops	Problems of identifying who is responsible for the pollution incident or outcome Problems of assigning monetary values to degradation of resource functions like atmospheric health or landscape beauty Unequal distributional impacts (such as of fuel taxes)

| Payments for environmental services | Based on Beneficiary Pays Principle – which holds that those who use and benefit from environmental services make payments to those who manage environmental resources to restore or establish land uses with external benefits | Creates markets for resource functions (especially regulating services such as carbon sequestration and water purification) currently not traded

Financial incentives for local actors to provide/maintain those services

Payments put monetary value on external benefits that thereby can become internal to farmers' and land managers' decisions | Voluntary transactions between buyers and sellers of environmental services

Substantial role for governments in establishing legal framework, tenure security, transparent monitoring, protecting indigenous people's rights etc. | UN Reduced Emissions from Deforestation and Forest Degradation (REDD+)

UN/World Bank Global Environment Facility National programme of Payments for Watershed Service in Costa Rica

Agro-environment schemes in many EU countries and the US | Many projects in early stages

Uncertain and contested environmental and human welfare benefits

Scientific uncertainty about links between land use decisions and environmental services and regarding rates of sequestration by vegetation type, for example

Problems of establishing baselines and additionality for projects |
| **Subsidies** | Reshape economic incentives to business to enable change | Payments made to firms and individuals to help them subsequently compete in open markets | Public monies used to support/incentivise private investment in activities of public benefit

Public authorities administer, accredit and license etc. | Widely used within renewable energy sector to support 'start-up costs' of new technology developments and/or to provide guarantees of future market access | Public funding may be withdrawn

Time lag between investment and realisation of benefits

Volatility of energy markets |

continued . . .

Figure 3.1 ... *continued*

Forms of governance	Conceptual basis	Market component	Role of public authorities	Examples	Issues
				'Feed in tariffs' used in over 60 countries to provide guaranteed purchase and prices for electricity generated from renewable energy sources	May require additional legal arrangements
				Many target small-scale generation on behalf of communities and individuals not traditionally engaged	
Voluntary/self-regulation	Individuals, groups, organisations take action to protect environment as good ecological citizens	Assumes action not for financial incentive May involve a willingness to pay extra for benign products and services	Not required by law Public authorities may provide infrastructure to support (recycling)	Fairtrade labelling Green consumerism Ethical investment Recycling Voluntary conservation work Voluntary carbon off-setting	Problems of scaling up from diverse, local successes Long-term success depends on internalisation of positive environmental attitudes and behaviour changes Concern as to whether the world can consume its way out of global environment and development problems

Source: compiled by author.

that sustainable development in practice requires. As seen in Chapter 2, the challenges of sustainable development include recognising the different values and interests surrounding environmental resources and functions, for example, and questioning where the costs and benefits fall, to whom, through particular development interventions. Sustainable development in practice cannot be achieved if processes of governance continue to create social and economic marginalisation, enhanced vulnerability and conflict. It will be seen through the chapter that improving 'governance' is now recognised by many development institutions as a critical agenda for more sustainable development.

International action

A principal means through which countries can confirm their cooperation within international efforts to support global environmental goals has been through their signatures to various 'multilateral environmental agreements' (MEAs). These are treaties that bind international behaviour towards collective objectives that could not be achieved by nation states acting individually. It is estimated that there are over 1,500 multilateral and bilateral environmental agreements currently in operation (Lemos and Agrawal, 2006). Table 3.1 highlights some of the most important in that they have the most signatories. Many MEAs have their origins in major international environmental summits. An average of 16 MEAs were signed each year between the Stockholm and Rio conferences (1972–1992) and 19 between the Rio and the World Summit on Sustainable Development in 2002 (Mitchell, 2003).

MEAs are not static documents; rather, they are renegotiated as parties to the agreement or circumstances change. This was seen in Box 2.6 in the case of the UN Framework Convention on Climate Change (UNFCC) agreed at Rio. At the time of signing it was what is termed 'soft-law' requiring voluntary commitments to reduce country emissions of carbon dioxide. Only with the full ratification of the Kyoto Protocol in 2004 did the convention become legally binding. However, the effectiveness of MEAs depends on states being willing to devolve some of their sovereign power to those created institutions (Werksman, 1996). Delays in the full ratification of the Kyoto Protocol were substantially due to the refusal of the US to sign on the basis that it would harm its domestic economy and that

Table 3.1 *Selected Multilateral Environmental Agreements (MEAs)*

MEA	Purpose	Date adopted	Entry into force	Number of signatories in 2010
Ramsar Convention – Convention on Wetlands of International Importance Especially as Waterfowl Habitat	To conserve and promote the wise use of wetlands.	1971	1975	70
World Heritage Convention – Convention Concerning the Protection of the World Cultural and Natural Heritage	To establish an effective system of identification, protection, and preservation of cultural and natural heritage, and to provide emergency and long-term protection of sites of value.	1972	1975	187
CITES – Convention on International Trade in Endangered Species of Wild Fauna and Flora	To ensure that international trade in wild plants and animal species does not threaten their survival in the wild, and specifically to protect endangered species from over-exploitation.	1973	1975	175
CMS – Convention on the Conservation of Migratory Species of Wild Animals	To conserve wild animal species that migrate across or outside national boundaries by developing species-specific agreements, providing protection for endangered species, conserving habitat, and undertaking cooperative research.	1979	1983	101
UNCLOS – United Nations Convention on the Law of the Seas	To establish comprehensive legal orders to promote peaceful use of the oceans and seas, equitable and efficient utilisation of their resources, and conservation of their living resources.	1982	1994	157
Vienna Convention – Convention for the Protection of the Ozone Layer	To protect human health and the environment from the effects of stratospheric ozone depletion by controlling human activities that harm the ozone layer and by cooperating in joint research.	1985	1988	188

Agreement	Objective			
Montreal Protocol – Protocol on Substances that Deplete the Ozone Layer (Protocol to Vienna Convention)	To reduce and eventually eliminate emissions of man-made ozone depleting substances.	1987	1989	196
Basel Convention – Convention on the Control of Transboundary Movements of Hazardous Wastes and Their Disposal	To ensure environmentally-sound management of hazardous wastes by minimising their generation, reducing their transboundary movement, and disposing of these wastes as close as possible to their source of generation.	1989	1992	176
UNFCCC – United Nations Framework Convention on Climate Change	To stabilise greenhouse gas concentrations in the atmosphere at a level preventing dangerous human-caused interference with the climate system.	1992	1994	195
CBD – Convention on Biological Diversity	To conserve biological diversity and promote its sustainable use, and to encourage the equitable sharing of the benefits arising out of the utilisation of genetic resources.	1992	1993	168
UNCCD – United Nations Convention to Combat Desertification	To combat desertification, particularly in Africa, in order to mitigate the effects of drought and ensure the long-term productivity of inhabited drylands.	1994	1996	194
Kyoto Protocol – Kyoto Protocol to the United Nations Framework Convention on Climate Change	To supplement the Framework Convention on Climate Change by establishing legally binding constraints on greenhouse gas emissions and encouraging economic and other incentives to reduce emissions.	1997	Not yet in force	193
Aarhus Convention – Convention on Access to Information, Public Participation in Decision-Making, and Access to Justice in Environmental Matters	To guarantee the rights of access to information, public participation in decision-making, and legal redress in environmental matters.	1998	2001	44

Source: World Resources Institute EarthTrends http://earthtrends.wri.org searchable database results.

it unfairly favoured developing countries. The latest discussions at Copenhagen of the post-Kyoto framework have confirmed the difficulty of such international 'rule making' that depend on all parties committing to an agreed course of action that may compromise national interests. Further difficulties arise from what is termed the 'democratic deficit', whereby countries participating may not be democracies, there is unequal distribution of power, knowledge and resources amongst participating countries and some countries are able to impose their preferences and undermine the capacity of other participants to impact on the final outcomes (Lemos and Agrawal, 2006).

MEAs also depend on the commitment of governments to ensure that measures are binding on their populations. Canada for example, has abandoned its commitments to their Kyoto targets on emissions reductions. Box 3.1 confirms the importance of national governments in ensuring the protection of environmental and human health in the context of the Basel Convention that seeks to control the trade in hazardous waste. Box 3.2 highlights that MEAs can, however, facilitate new partnerships such as between governments and business. It is also an illustration of 'cap and trade' as a mechanism to deliver environmental improvements. Despite substantial debate regarding the future of such 'state-centred' institutional arrangements, it is also considered that whilst MEAs are not perfect, 'the world would be a much poorer and endangered place if such international agreements did not exist at all' (O'Riordan, 2009: 312).

Box 3.1

Controlling the trade in hazardous waste

In 1989, the Basel Convention on the Control of Transboundary Movements of Hazardous Wastes and Their Disposal was developed as a multilateral environmental agreement to address the emergent problem of firms (largely from the US and Europe) 'dumping' hazardous materials (defined as waste which, if deposited into landfills, air or water in untreated form will be detrimental to human health or the environment) into unprotected communities in Africa, the Caribbean and Latin America; an emerging trade that was understood as being related to the rising costs of disposal within the source countries associated with tightening of domestic environmental regulations. The convention aims to establish a framework for sound environmental management of hazardous waste through the minimisation of both the generation and transboundary movement of such wastes and to ensure disposal as close to the source of production as possible. Where wastes were to be exported, the exporting country must have 'prior informed consent', that is, written consent from an appropriate authority in the receiving country with an obligation on the exporter to ensure that the wastes at destination would be managed in an environmentally sound manner. The only acceptable justification for hazardous waste export under the convention is if a country lacks adequate technical capacity to handle/manage those wastes domestically or the importing country requires the waste as a raw material. An amendment to the convention in 1994 calls for all OECD countries to *ban* export of hazardous wastes to non-OECD countries.

Whilst the Basel Convention, along with a number of regional agreements (such as the Bamako Convention of 1991 which bans the imports of hazardous wastes into Africa) and national legislation (including within industrialising countries such as China and India) have served to limit the most obvious cases of the export of hazardous waste in recent years, the international trade has not been eliminated. For example, in August 2006, the movements of a Panamanian registered ship, the *Probo Koala* (chartered by a British-owned oil and commodity shipping company, Trafigura) attracted international media attention. The focus of concern was the disposal of its waste cargo in a number of locations around the city of Abidjan, the capital of the Ivory Coast and the subsequent claim by thousands of local people that they had fallen ill as a result.

Before arriving in Abidjan, the ship had already been turned away from the Dutch port of Amsterdam where local residents reported the strong smell of 'rotten eggs' (hydrogen sulphide) and the Port Authorities had established the toxic nature of the cargo. A Dutch company tendered to dispose of the waste for 500,000 Euros. However, days later the *Probo Koala* departed for the Ivory Coast where a local company (allegedly registered only days before the arrival of the ship and involving senior civil servants) was contracted to dispose of the waste at the substantially reduced cost of 18,500 Euros. In the weeks subsequent to the offloading of the cargo, as many as 17 people had died and over 40,000 people had sought medical treatment for effects including headaches, stomach pains, respiratory problems and vomiting (according to news media in the UK and US).

The case has prompted high-profile attention including:

● The nine-month-old transitional government in Ivory Coast resigned in September 2006
● Independent inquiries were launched by both the city of Amsterdam and the United Nations
● A class action was raised on behalf of 31,000 local Ivorians by a UK firm in the High Court in London (the largest injury case in British history). In 2009 an out-of-court settlement of £30 million was agreed.

However, there remain questions concerning what was on board the ship, the ports visited and who could be held responsible in future. Trafigura asserts that the cargo was 'slops' from routine washing of tanks, comprising a mixture of water, gasoline, caustic soda and a small amount of hydrogen sulphide. Other suggestions are that the cargo was a consignment of coker gasoline (bought from a Mexican, state-owned oil company) that was subjected to a process of onboard 'caustic washing' and resold for a reported profit of US$19 million. On this basis, Trafigura was looking to dispose of a highly toxic waste including sodium hydroxide, sodium sulphide and phenols. Trafigura continues to deny responsibility and suggests that it could not have foreseen the failure of the local company to dispose of the slops in an appropriate manner (it was allegedly spread over a number of public dumps, waste ground and roadsides).

However, as signatories to the Basel Convention and under legally binding EU law, it is also possible that EU countries visited by the *Probo Koala* could be liable to take the waste back if they were a country of export. Evidence suggests, for example, that the Dutch authorities had concerns about the waste on the ship and would have been obligated under the Convention to prevent the ship from leaving with the waste.

Source: compiled from Evans (2010), Verkaik (2010).

Box 3.2

The contribution of the Clean Development Mechanism to sustainable development

The Kyoto Protocol became legally binding on 'Annex 1' countries in 2004. It set an 'environmental ceiling' or 'cap' for future global greenhouse gas emissions of 5.2 per cent below 1990 levels by 2012. A key part (and entirely new principle) of the Protocol was that some emissions reductions required of countries could be met by transferring to other counties ('trading' emissions permits) under what are known as Flexibility Mechanisms. The underpinning rationale of these mechanisms was that enabling a cross-border trade in emissions reductions would facilitate cheaper reductions in emissions globally. Less developed economies (who were not subject to the binding targets on emissions reductions) could sell 'carbon credits' or so-called 'certified emissions reductions' to countries (and industry) that did have these emissions commitments. As a result, emerging economies would be supported to make energy savings and to invest in less polluting technologies. One of the Flexibility Mechanisms was designed to generate new project-based credits whereby a company or investor from a developed country can fund a project in a developing country that either reduces emissions or enhances sinks. This is known as the Clean Development Mechanism (CDM). Such projects must produce reductions that are additional to any that would occur in the absence of certified project activity. Their explicit purpose stated in the Protocol was that they should assist Parties not included in Annex 1 in achieving sustainable development as well as contribute to the overall global objective of lowering emissions.

There is much debate as to the success of the CDM to date. One concern is the inequitable distribution of projects across developing regions: 73 per cent of investment/trading in terms of value has been with China, where costs of emissions abatement have been low (Boyd et al., 2009). Yet China is one of the fastest growing economies in the world (as well as the largest current greenhouse gas emitter). In contrast, countries across sub-Saharan Africa have less than 3 per cent of all projects registered. Furthermore, whilst it was intended that the CDM would stimulate investment in renewable technologies such as wind, solar and geo-thermal energies, the majority of projects to date have been in other sectors; in gas capture projects at major chemical and manufacture plants addressing emissions of HFC and N_2O, for example, that are in fact more 'minor' greenhouse gases in terms of their contribution to climate warming than CO_2. Critics question whether companies should be being 'paid' to clean up a mess of their own making. From an environmental justice perspective, the concern is that CDM projects are being used to provide generous subsidies to companies to use existing technologies to mop up industrial gases rather than shifting to the low carbon world that is needed.

There are also concerns as to the contribution of CDM projects to sustainable development. The process of identifying projects for the CDM include the host government (through its Designated National Authority) being responsible for defining the criteria for sustainable development in any project context. Project proposals are then passed through a UN body for validation. However, an analysis of 65 Indian project documents for their stated contribution to sustainable development concluded that they 'offer just lip service regarding expected contribution to socioeconomic development of the masses, particularly in rural areas' (cited in Boyd et al., 2009: 822). Whilst many projects to date have made significant emissions reductions, they are falling short in delivering local benefits, either directly such as through employment or indirectly through an improvement in local environmental and social conditions. Furthermore, there is little evidence for the participation of civil society in the decisions regarding projects, leading to enhanced local tensions where projects serve to support business and industry that are in fact causing some of the worst social and environmental problems locally. It is acknowledged that the CDM can be a useful approach to encourage the development of emissions-reduction projects in developing countries, but there are a number of issues to be addressed as part of reforming the CDM that is part of post-Kyoto negotiations.

Aid and finance for the environment

Foreign aid is defined as any flow of capital to the developing nations which meets two criteria. First, its objective should be non-commercial from the point of view of the donor; second, it should be characterised by interest and repayment terms which are less stringent than those of the commercial world. The notion of foreign aid is that these grants and loans transfer financial resources from more wealthy to poorer nations to promote development and welfare. Historically, what is termed 'official development assistance' (ODA) has comprised grants and concessional loans from individual governments (traditionally from higher-income countries and in particular members of the OECD) and through the multilateral development institutions (such as the World Bank). In recent years, further aid channels have proliferated to include new sources of ODA (the governments of India and Korea are now donors despite also being recipients of aid, for example), an expansion of private organisations (especially from the US and including the Melinda and Bill Gates Foundation and Ford Foundation) and increased funding through NGOs. In addition, new networks combining bilateral ('government to government'), private, NGO and finance through UN organisations are also being seen around specific issues – notably HIV/AIDS and malaria. Figure 3.2 confirms the many sources of ODA currently for Africa. In many ways, aid is now a complex business. Deutscher and Fyson (2008) for example suggest that there are more than 280 bilateral donor agencies, 242 multilateral programmes, 24 development banks and about 40 United Nations agencies now involved and an estimated 340,000 development projects around the world.

There are long-standing debates over the impact of aid on recipient nations. Opinions range from the belief that it is an essential

Figure 3.2 *Percentage share of Africa's ODA budget*

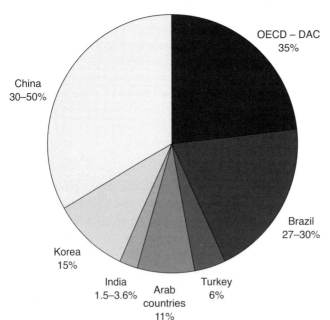

Source: Africa Progress Panel (2010).

prerequisite for development (supplementing scarce domestic resources) to the view that aid perpetuates neo-colonial and dependent relationships that prevent development processes including the spread of democracy. Many of these arguments have been substantially reinvigorated in the current context, in particular the proliferation of aid sources, the proximity of the MDG target deadlines (2015), the global economic crisis and the recognised challenges of finding new finances to deliver on climate commitments. Whilst Dambisa Moyo argues for the halting of aid, suggesting that 'aid has been and continues to be, an unmitigated political, economic and humanitarian disaster for most parts of the developing world' (Moyo, 2009: xix), the UK government 'ring-fenced' its aid budget in 2009 to ensure that it can deliver on its existing commitments at a time of severe cuts being made in domestic social spending. Many authors and institutions call for 'less and better aid' (see Glennie, 2008).

As far back as 1970, the UN set targets for ODA, urging developed countries to allocate 0.7 per cent of GNP annually by 1975. This target been repeatedly endorsed at international development conferences yet has only been reached by a few countries, as seen in Figure 3.3. The US has never signed up to the 0.7 per cent target, but it is estimated that half of all private philanthropic aid to developing countries comes from the US (Curto, 2007). Calls for enhancing levels of aid have been made, for example at the G8 meeting in Gleneagles in 2005. 'Scaling up' aid is recognised as important for meeting the MDG commitments and in relation to the particular needs of the African continent. However, there remain concerns as to whether commitments made at such summits are realised in practice, arrive 'on time' and are 'additional' to existing promises (all of which are essential for planning within host countries, for example). The Gleneagles commitment to provide a further US$25 billion to African countries by 2010 in fact realised only US$11 billion (Africa Progress Panel, 2010).

Rather than finances going directly to governments, aid now flows through many different channels of disbursement, evaluation and often staffing. Figure 3.4 identifies a number of problems associated with these new trends. An overarching concern is the impact on how development decisions are now made within a country and the capacity for host countries and people to set their own policy and priorities. Indeed, Brazil, Argentina and Bolivia have recently turned down IMF loans under pressure from civil society groups concerned

Figure 3.3 *ODA as proportion of Gross National Income, 2009*

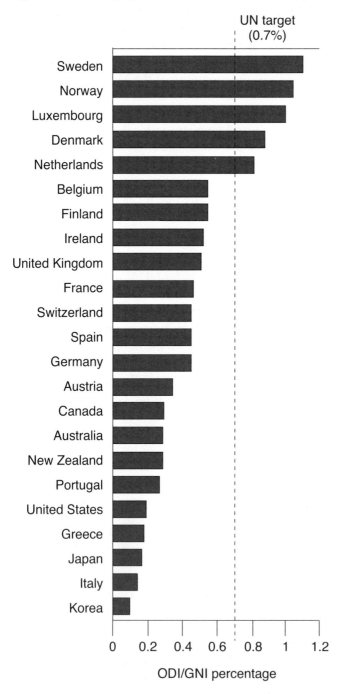

Source: adapted from OECD (2010a).

that such lending in the past has distorted domestic economic and social policy. The constant reporting to donors that is required may also compromise the prospects for developing stronger governance structures within these host countries. The Monterrey Consensus signed by members of the UN in 2003 is one attempt to enhance the coherence, consistency, effectiveness and transparency of aid in future.

Figure 3.4 *The challenges of aid*

Aid fragmentation:

In 2005–06, 38 developing countries received ODA from more than 25 DAC and multilateral donors. In 24 of these countries, 15 or more donors collectively provided <10 per cent of that country's total aid (Deutscher and Fyson, 2008: 16).

The average number of donors per country was 12 in the 1960s. It is now 33 (Glennie, 2008).

Aid coordination:

In 2005, Vietnam received 791 missions from donors – more than two a day including weekends and holidays (Deutscher and Fyson, 2008: 16).

In 2006, Tanzania received 541 donor monitoring missions (Glennie, 2008).

Aid priorities:

In Rwanda, 75 per cent of ODA goes directly through NGOs and/or donor-managed projects. Malaria is the leading cause of mortality in Rwanda, but donor funds to anti-malaria activities are one-third of those to HIV/AIDs (Curto, 2007).

'Even India is giving aid to Africa . . . with 400 million poor people of its own to attend to, this largesse is obviously not about poverty reduction' (Glennnie, 2008: 111).

The burden of monitoring:

Most African countries submit approximately 10,000 quarterly donor reports every year (Glennie, 2008).

The government of Tanzania declares a donor-mission-free period in each year such that civil servants can get on with their main jobs (Glennie, 2008).

The Monterrey Consensus:

This urges members of the UN to intensify their efforts as aid donors to:
● harmonise their operational procedures
● move to further untie aid and remove restrictions
● enhance capacity of recipient countries
● use frameworks that are owned and driven by developing countries
● enhance recipient country input to design of technical assistance programmes
● improve ODA targeting to poor.

Aid and environment

Certainly, there is much evidence that aid can be (and has been) environmentally damaging. In Chapter 1 it was seen that the widespread damage to environments and local peoples caused by resettlement projects, large dams and road building, particularly in Latin America (and those financed by World Bank), were important in shaping public environmental concern in the US in the 1970s. Concern regarding the negative environmental impacts of structural adjustment programmes (SAPs) was also identified as the focus for emerging NGO campaigns and actions both within developing countries and transnationally. SAPs were the conditions required of countries to access funding from the World Bank/IMF, but also to be eligible for debt relief (which 'counts' for donor countries as 'ODA'). They have been important therefore in releasing domestic financial resources for investments in environment and development. The section below considers the impact of debt (and measures to address it) on the environment more fully. As seen above, more ODA is now disbursed through NGOs and one reason is the suggested characteristics of NGOs that make them better suited to effect more sustainable development. This is explored further in the final section of this chapter.

It is only recently that ODA commitments for environmental outcomes have been monitored. Whilst there are acknowledged problems of accounting (such as whether finances go to explicitly environmental projects or to those that have been modified in small ways to support those objectives), approximately 30 per cent of bilateral aid in 2007/8 was considered in support of the environment (OECD, 2010b). The new commitments being made on climate finance are also classified as ODA. In 2008, approximately 8 per cent of total ODA (US$8.5 billion) was aimed at climate change mitigation, that is, supporting recipient countries to reduce emissions, enhance carbon sinks and/or integrate climate change concerns into their development objectives (OECD, 2010c). Finance for climate change adaptation is largely through the Adaptation Fund set up in 2007 under the Kyoto Protocol. This fund uses finance raised through the CDM in particular (Box 3.2) as well as other donations to support measures to anticipate and respond to climate change impacts. Most recently, under the 2009 Copenhagen Accord, developed countries have committed to provide US$100 billion per year to developing countries by 2020 to help responses to climate change. It is unclear at present if these commitments will be as loans or grants and whether they will be 'new and additional'. A concern is that existing aid commitments could be renamed and therefore diverted from other important development/environment objectives.

Box 3.3 identifies a scheme that has attracted much interest from many different donor organisations as a means for raising finances for the environment and for disbursing their aid commitments under the Copenhagen Accord. It is an illustration of a 'market mechanism' as identified in Figure 3.1 to deliver environmental objectives that rests on payments being made by the global North to communities and governments in the South in respect to their commitments to protect global carbon sinks (particularly forests).

Box 3.3

Reducing Emissions from Deforestation and Forest Degradation (REDD+)

In 2007, the parties to the UNFCC committed to place efforts to reduce emissions from deforestation and forest degradation in developing countries (REDD+) at the centre of discussions regarding any new climate agreement when the current Kyoto agreement ends in 2012. In short, the existing Kyoto Protocol has focused on responsibilities and targets for GG emissions but has not accounted for any actions to support and conserve global carbon sinks and in particular forests. Whilst the nature of the post-Kyoto international agreements on climate remain uncertain, the Copenhagen Accord indicates continued support for REDD+. Wider political and stakeholder interest in REDD+ is also high from, among others, the World Bank through its Forest Carbon Partnership Facility, governments of the EU, the Coalition of Rainforest Nations (Costa Rica and Papua New Guinea are proactive proponents) and plantation, oil and gas companies.

The essential idea behind REDD+ is that countries, companies, communities or individuals could be paid to manage forests and reduce deforestation in recognition of the role forestry has in carbon sequestration and the significance of land use change in annual carbon emissions at a global scale. The beneficiaries of such actions – the global community – would pay forest owners for the services they are providing (maintaining global atmospheric systems and all that depend on those) and compensate them for any additional costs and/or loss of income that such longer-term conservation ('avoided deforestation') requires.

The principal of REDD+ acknowledges that powerful economic incentives are needed to discourage forest clearance given the strength of current economic drivers of deforestation such as palm oil expansion, conversion to agriculture and industrial tree planting. It is anticipated that funding could come either from aid money and/or through the carbon market (the latter being preferred by the World Bank and EU governments). The UN suggests that REDD+ could generate a North–South flow of finances of US$30 billion per year (cited in FoE, 2010: 5). As such REDD+ can be considered a global-scale 'payments for environmental services' mechanism, as considered in Figure 3.1.

Whilst there is currently no international agreement on REDD+, there are several large scale initiatives including the UN-REDD programme that are already supporting countries in capacity building, strategy development and pilot projects in preparation to implement REDD+. 'Interim' activities are underway in 54 developing countries (FoE, 2010) with the majority funded through bilateral aid. Norway, for example, is currently in collaboration with both Indonesia and Brazil for projects worth US$1 billion in each (FoE, 2010: 11). However, there are fears that without international agreement, decisions will be donor driven and fragmented. In addition, such top-down policy driven by global agencies and governments could have serious negative impacts, as seen in Figure 3.5, on forest-dependent communities and indigenous peoples whose culture and livelihoods are closely linked to those forests. There is also the overarching ethical concern regarding carbon markets, where industrial and corporate interests pay for carbon credits to be used within schemes such as REDD+ whilst continuing to pollute and extract fossil fuels elsewhere (potentially at a cost to other people). Further problems include the fact that REDD+ does not address the underlying drivers of deforestation to reduce demand for timber and agricultural commodities or the fundamental challenge of creating lower carbon economies to mitigate climate change. A suggestion is that REDD+ could become about creating new carbon markets and opportunities to make money.

Figure 3.5 *The potential risks to livelihood of avoided deforestation policies*

- Increased state and 'expert' control over forests
- Overzealous government support for anti-people models of forest conservation (evictions and expropriation) with forests seen as lucrative carbon reservoirs
- Unjust targeting of indigenous and marginal peoples as the drivers of deforestation
- Violations of customary and territorial rights
- State and NGO zoning of forests without informed participation of forest dwellers
- Unequal imposition of the costs of forest protection on local communities
- Unequal and abusive community contracts
- Land speculation, grabbing and conflicts
- Embezzlement of international funds by national elites
- Increasing inequality between those in receipt of funds and those who are not
- Potential conflict within communities over acceptance or rejection of schemes.

Source: compiled from Griffiths (2007).

The World Bank in sustainable development

The 'World Bank' (WB) comprises a number of institutions, as shown in Figure 3.6. The origin of the World Bank (and its 'sister' organisation, the International Monetary Fund) is the Bretton Woods Conference of 1944, and was part of a framework aimed at ensuring the reconstruction and development of Europe after the Second World War. In operation, since 1950 the World Bank has loaned monies increasingly to the governments of developing nations. To qualify for WB loans, countries must first be members of the IMF (which is concerned with economic and financial stability and sets and oversees codes of economic conduct on behalf of its members). The WB is the major source of multilateral finance for developing countries. In 2010, gross disbursements from IBRD and IDA totalled US$58.7 billion (World Bank, 2010c). In addition, it is known that WB lending decisions influence what other private banks and donor organisations do, such that for each dollar that the World Bank lends, it can be expected that many more will also flow to these projects and initiatives (Rich, 1994). The rhetoric and actions of the World Bank with regard to the environment are therefore crucial in determining the prospects for sustainable development.

Fundamentally, the WB, as a multilateral institution, has been able to borrow money on world markets and lend more cheaply than

Figure 3.6 *The World Bank Institutions*

The International Bank for Reconstruction and Development (IBRD)
● Established 1944
● 187 members
● Lends to governments of middle-income and creditworthy low-income countries
● Lending in 2010: US$44.2 billion for 164 new operations in 46 countries

The International Development Association (IDA)
● Established 1960
● 170 members
● Provides interest-free, long-term loans – called credits – and grants to governments of the world's 79 poorest countries (defined in 2011 as per capita incomes of <US$1,165)
● Lending in 2010: $14.5 billion for 190 new operations in 66 countries

The World Bank Group also includes the international Finance Corporation (IFC), the Multilateral Investment Guarantee Agency (MIGA) and the International Centre for Settlement of Investment disputes (ICSID).

Source: compiled from World Bank (2010c).

commercial banks. It raises money by selling bonds and other securities to individuals, corporations, pension funds and other banks around the world. Its securities are considered to be amongst the worlds safest, ensuring that it can borrow on very favourable terms and is able to lend money at rates of interest below commercial levels. Decisions on allocating funds take place on the basis of 'one dollar–one vote', that is, according to the financial contribution of the voting country to the bank. On this basis, approximately 45 per cent of voting rights are held by the G8 countries, with developing countries typically having less than 0.1 per cent of votes. In the early decades, World Bank lending was principally to projects, that is, investments in discrete, time-bound and often large-scale initiatives such as within infrastructure and mining. From the early 1980s, an increasing proportion of lending has been 'programme lending', for packages of policy reform to address the failings in the macro-economies of developing countries that were compromising what could be achieved through traditional project lending. As seen in Chapter 1, SAPs were considered by the World Bank and IMF as key to the longer-term solution of the debt crisis. By the end of the 1990s, over 50 per cent of World Bank lending was to packages of macro-economic and policy reform (Potter et al., 2008).

Figure 3.7 identifies the countries that receive World Bank financial support. Table 3.2 shows the destinations of these monies in terms of particular development themes, the patterns of which are discussed further below. The World Bank also provides a number of 'non-lending services' to developing countries. These include technical and policy advice and research (both at a general level, as in the annual *World Development Reports*, and within countries to support environmental and poverty assessments, for example). It also provides support for consultations and impact evaluations.

Table 3.2 *World Bank lending by theme, 2010*

Theme	Percentage of total lending
Financial and private sector development	30
Human development	14
Public sector governance	10
Rural development	9
Urban development	9
Social protection and risk management	9
Economic management	7
Environment and natural resource management	7
Trade and integration	3
Social development, gender and inclusion	2
Rule of law	<1

Source: adapted from World Bank (2010c).

Figure 3.7 *Countries receiving financial support from the World Bank*

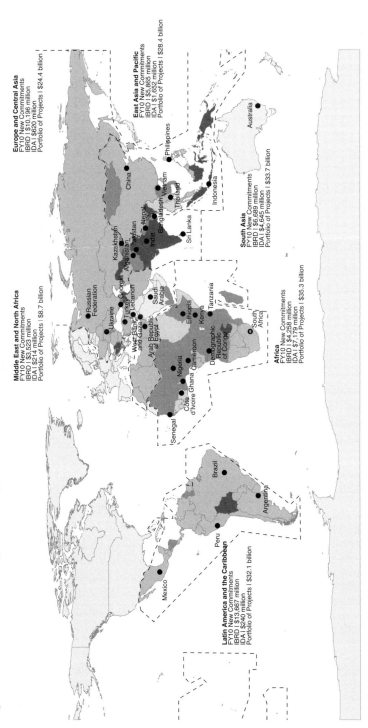

Europe and Central Asia
FY10 New Commitments
IBRD I $10,196 million
IDA I $620 million
Portfolio of Projects I $24.4 billion

East Asia and Pacific
FY10 New Commitments
IBRD I $5,865 million
IDA I $1,652 million
Portfolio of Projects I $28.4 billion

Middle East and North Africa
FY10 New Commitments
IBRD I $3,523 million
IDA I $214 million
Portfolio of Projects I $8.7 billion

South Asia
FY10 New Commitments
IBRD I $6,689 million
IDA I $4,645 million
Portfolio of Projects I $33.7 billion

Africa
FY10 New Commitments
IBRD I $4,258 million
IDA I $7,179 million
Portfolio of Projects I $35.3 billion

Latin America and the Caribbean
FY10 New Commitments
IBRD I $13,667 million
IDA I $240 million
Portfolio of Projects I $32.1 billion

Countries eligible for IBRD funds only
Countries eligible for a blend of IBRD and IDA funds
Countries eligible for IDA funds only
Inactive IDA-eligible countries
Countries not receiving World Bank funds
● Offices with the Country Director present
– – World Bank Regions

Russian Federation, Ukraine, Turkey, Georgia, Lebanon, West Bank and Gaza, Arab Republic of Egypt, Saudi Arabia, Kazakhstan, Afghanistan, Nepal, India, Bangladesh, Sri Lanka, China, Vietnam, Thailand, Philippines, Indonesia, Australia, Senegal, Côte d'Ivoire, Ghana, Nigeria, Cameroon, Democratic Republic of Congo, Ethiopia, Kenya, Tanzania, South Africa, Mexico, Peru, Brazil, Argentina

Source: adapted from World Bank (2010c).

The World Bank certainly has its opponents and as Nelson (2008: 552) suggests, has regularly been a 'lightning rod for criticisms of the international economic system and of development aid'. It has been and remains a key target of environmental critics. A recent independent evaluation, however, has suggested that World Bank support for the environment has evolved from a preventative approach of 'do no harm' to a more proactive 'do good' approach in recent years (World Bank, 2008a). The following sections consider the evidence for changes in the World Bank operations in relation to the environment, in terms of its investment and policy lending and in the way in which it works with other institutions towards sustainable development.

Greener project lending

Early environmental activities of the World Bank were largely concerned with assessing and mitigating the environmental impacts of its lending. An Office of Environmental Affairs was created in 1973 for this purpose. However, it had minimal staff and it was not until 1987 that a central Environment Division was established to promote environmental activities (as well as oversee them). In 1992, a 'four-fold environmental agenda' (Figure 3.8) for the World Bank was announced reflecting the agreements reached at the Earth Summit. The World Bank committed to supporting countries to develop National Environmental Action Plans, to assisting the integration of the interdependent concerns of poverty reduction, economic growth and environmental protection and to administer the new finances of the Global Environment Facility (all key outcomes of the Rio process as seen in Chapter 1). Commitments were also made to improve Environmental Impact Assessment (EIA) and 'safeguard' policies (such as those referring to indigenous peoples) within project lending.

The seriousness of the World Bank's commitments to environmental reforms was quickly tested in the early 1990s. The USA

Figure 3.8 *The four-fold environmental agenda of the World Bank*

1. Assisting member countries in setting priorities, building institutions and implementing programmes for sound environmental stewardship.
2. Ensuring that potential adverse environmental impacts from bank-financed projects are addressed.
3. Assisting member countries in building on the synergies among poverty reduction, economic efficiency and environmental protection.
4. Addressing global environmental challenges through participation in the Global Environment Facility (GEF).

Source: World Bank (1994).

(the WB's largest contributor) threatened to withhold 25 per cent of its 1992 contribution (approximately US$70 million) in relation to bank involvement in the damming of the Narmada River in India. At the time, this was the world's largest hydroelectric and irrigation complex, based on 30 major, 135 medium and 3,000 minor dams to be built over 50 years. It was designed to generate an estimated 500 million megawatts of electricity, irrigate over two million hectares and bring drinking water to thousands of villages. The scheme, however, was mired in controversy from the outset: the dams would displace 200,000 people, submerge 2,000 square kilmetres of fertile land and 1,500 square kilometres of prime teak and sal forest, and eliminate historic sites and rare wildlife. Fierce local and international protests led to the WB taking the unprecedented step of commissioning an independent review of its activities and subsequent support was withdrawn. Whilst the Indian government continued with the project, civil society challenges from within the country persisted, including petitions of the Narmada People's Movement to the Supreme Court of India (Vidal, 1998).

Strengthening of EIA procedures continues to be key to processes to reduce the negative environmental impacts of World Bank project lending. Before projects are approved for lending, they are 'screened' for prospective environmental and social impacts and various levels of investigation (including public consultations) are triggered before the project can proceed. However, a recent review has suggested continued problems of non-compliance with the Bank's environmental assessment policy (World Bank, 2008) and weaknesses in those assessments that are undertaken (particularly regarding public consultation and in determining indirect and cumulative environmental impacts).

A further element of the World Bank's environmental agenda has been to provide funds for projects which specifically aim to strengthen environment and natural resource management (termed ENRM lending and identified in Table 3.2). Such projects include direct investment in pollution prevention and treatment, the conservation of biodiversity, integrated river management and establishing national parks as well as funds for research, capacity building, training and monitoring. Between 1990 and 2007, approximately 15 per cent of all World Bank commitments (by amount of lending) included ENRM components (World Bank, 2008a). Table 3.3 shows the regional destination of this lending. East Asia is the largest recipient and 57 per cent of these

Table 3.3 ENRM lending by region, 1990–2007

Region	Percentage of lending
East Asia and Pacific	33.3
Latin America and Caribbean	19.4
South Asia	14.8
Europe and Central Asia	14.1
Sub-Saharan Africa	11.4
Middle East and North Africa	6.9

Source: adapted from World Bank (2008a).

commitments are to China specifically. India receives more such funding than the whole of sub-Saharan Africa (World Bank, 2008: 23). The majority of ENRM commitments (95 per cent by value) are within IBRD/IDA projects. However, they also include the new finances for climate change identified above and those to the Global Environment Facility (GEF).

The Global Environment Facility was established in 1990 in conjunction with UNEP and UNDP as a programme of new monies to assist the least developed nations in the arenas of climate change, biodiversity protection, international water management and ozone depletion. Monies were to help meet the additional costs of transforming a project with national benefits to produce global benefits (or indeed 'witholding development' to protect the global environment). Through GEF, the World Bank currently operates in partnership with ten UN agencies, numerous NGOs and the private sector to fund approximately 1,800 projects (the costs ranging from US$2,000 to $50 million) in over 140 countries (Mee et al., 2008). GEF is the main funding mechanism for delivering the goals of the Convention on Biological Diversity and the Framework Convention for Climate Change. GEF projects, by thematic area since its inception, are shown in Figure 3.9. It has been praised by some as an important tool for multilateral cooperation on environmental issues, but is criticised by others for focusing on what could be considered 'western' environmental agendas (Werksman, 1995) and being too closely aligned to World Bank decision-making and projects (Jordan and Brown, 2007). Since 2008, GEF has been the principal channel for international commitments to the WB's Climate Investment Fund. However, there is some concern that linking climate funds to GEF (with its established priorities to maximise *global* environmental benefits) may lead to smaller countries being unable to access these finances in future (World Bank, 2008a: 22).

Programme lending and sustainable development

Structural adjustment programmes were the key mechanism used to address the debt crisis from the late 1980s. However, concern

Figure 3.9 *Distribution of GEF funding by focal area, 1991–2010*

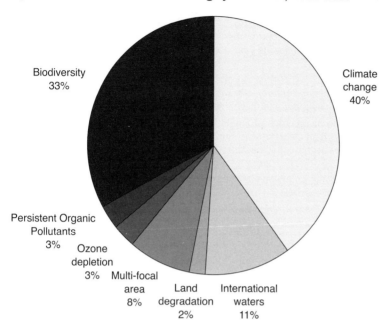

Note: Total disbursement $4.5 billion

Source: http://go.worldbank.org/JBZVXOC7OO, accessed 23 February 2012.

quickly mounted that SAPs were undermining the prospects of sustainable development (Reed, 1996; FOE, 1999). Mohan et al. (2000) suggested that SAPs were 'environmentally blind' in the first ten years of implementation. The focus was generally on fiscal issues (i.e. the balance of payments difficulties and governments spending more than they had) and the environment was generally not high on government or public agendas. Fundamentally, the core of neo-liberal thinking encapsulated in SAPs was for decreasing state expenditure, not increasing it through environmental protection. 'In the effort to rapidly trim budget deficits, governments are forced to make choices, and inevitably, the environment loses' (FOE, 1999: 4). Not only were the budgets and staffing of environment departments regularly cut, weakening the ability to enforce environmental laws, but environmental legislation may have been relaxed in order to increase foreign investment (FOE, 1999). Figure 3.10 illustrates some of the environmental outcomes of these pressures.

It was also quickly evident that structural reform was widening socioeconomic and gender disparities. The UNICEF publication of 1987 *Adjustment with a Human Face* was particularly important for

Figure 3.10 *Pressures of adjustment on the environment*

- In Cameroon, the IMF-recommended export tax cuts and devaluation of the currency in 1995 led to increased incentive to export timber. The number of logging enterprises rose from 194 in 1994 to 351 in the following year. Exports of lumber grew by 50% between 1995 and 1997.
- Under SAP guidance since the mid-1980s, Guyana has implemented policies to increase large-scale, foreign-owned mining ventures. There are now 32 foreign mining companies active in the country with mining permits, covering 10% of the country.
- SAP in Tanzania resulted in rising costs of inputs for agriculture. Production increases were pursued through increased land clearing at a rate of 400,000 hectares per year. Between 1980 and 1993, a quarter of the country's forest area was lost (40% of which was to cultivation).
- In Brazil, government spending on environmental programmes was cut by two-thirds in order to meet the fiscal targets of the IMF.
- Benin, Guinea, Mali and the Central Africa Republic all established new mining codes to promote exploration and development.

Source: compiled from Friends of the Earth (1999).

exposing the increased poverty and social polarisation (particularly the worsening fate of women and children under five) occurring under adjustment. It also revealed the pressure on natural resources being created by people's coping strategies under such conditions. Through such evidence (and the wider resistance to neo-liberalism and globalisation considered in Chapter 1), pressure mounted on the World Bank and other financial institutions to modify the conditions of their policy lending. Targeted funding, for example, became available for various 'Social Safety Nets' aimed to reduce what were considered short-term, negative impacts of adjustment on certain groups (for an overview and critique of their impacts, see Potter et al., 2008).

In 1999, SAPs were replaced by Poverty Reduction Strategy Papers (PRSPs). The intention is that PRSPs could overcome many of the limitations of SAPs. For example, SAPs were characterised by very prescriptive conditions on recipient governments, had tended to embrace a 'one size fits all countries' approach (see Figure 1.11) and public citizens or NGOs had been excluded from the processes involved. PRSPs are developed by individual national governments and through much more explicit and open participatory processes that include representatives of the World Bank and IMF but also other donors and civil society, as seen in Figure 3.11. PRSPs are required to set out coherent plans for reform focused on poverty reduction and to identify their financing needs in respect to their particular social, economic and environmental contexts, that is, are tailored to the

Figure 3.11 *The core principles of the PRSP approach*

Poverty reduction strategies should be:

- **Country driven**, promoting national ownership of strategies through broad-based participation of civil society
- **Result-oriented** and focused on outcomes that will benefit the poor
- **Comprehensive** in recognising the multidimensional nature of poverty
- **Partnership-oriented**, involving coordinated participation of development partners (governments, domestic stakeholders, and external donors) and
- Based on a **long-term perspective** for poverty reduction.

Source: IMF (2010).

specific conditions and needs of particular countries as defined by multiple stakeholders. The World Bank also 'encourages' governments to consider environmental factors in their PRSPs, 'because of the links between environment and poverty, and because a poverty reduction strategy must be environmentally sustainable over the long term' (World Bank, 2001a: 144).

An analysis of 40 countries implementing PRSPs to 2005 suggests that this approach has made important progress in putting poverty as a stronger focus within government, engaging civil society to an unprecedented extent in policy activities and enhancing coordination of donors at a country level (Driscoll and Evans, 2005). However, there is also considerable variation across countries in the extent to which the environment is integrated. In a review of 53 PRSPs, Bojo et al. (2004) looked at mainstreaming of the environment in terms of: how environmental issues were explained; any analysis of poverty–environment linkages; identification of environmentally relevant actions; the extent to which participation and consultation allow environmental concerns to be heard; and any alignment with MDG7 (on sustainable development). Whilst some PRSPs had very thorough engagements, in others it was only marginal. The authors identified that environmental health issues generally got more attention than natural resource management issues and only a few took a longer-term perspective. Only 14 had any explicit targets aligned to MDG7 and in most cases this was focused on the water and sanitation target.

Strategic development and change within the World Bank

In 2001, the World Bank launched its first environment strategy, *Making Sustainable Commitments*. The aim was to link environment and development throughout all WB activities in future through three core objectives, identified in Figure 3.12. Importantly, there was recognition that change was required in the bank itself: to 'accelerate the shift from viewing the environment as a separate, freestanding

Figure 3.12 *Strategic objectives for sustainable development at the World Bank*

- Improving the quality of life (focusing on three broad areas where environment, quality of life and poverty reduction are strongly interlinked – in enhancing livelihoods, preventing and reducing environmental health risks and reducing people's vulnerability to hazards)
- Improving the quality of growth and in particular ensuring that short-term gains do not come at the expense of constrained opportunities for future development
- Protecting the quality of the regional and global commons in recognition that solutions to sustainability need to go beyond individual countries

Source: World Bank, (2001b).

concern to considering it an integral part of our development assistance' (World Bank, 2001b: xxv).

The Bank is currently undertaking consultations on a new Environment Strategy. It recognises that the external context has changed considerably since the first strategy, in particular the impact of climate change on development gains in its client countries and the pace and extent of urbanisation (World Bank, 2009c). In 2008, an Independent Evaluation Group assessment of World Bank support to the environment (World Bank, 2008a) had found that whilst support to National Environmental Action Plans and general capacity building for environmental stewardship had improved, this had not been followed through in project lending. The review also identified problems in the performance of ENRM projects within lower-income countries and especially in sub-Saharan Africa. Similarly, whilst much analytical work had been done at a general and country level regarding the linkages between poverty, livelihoods and environment, the review found that much lending (particularly policy-based lending) still gave insufficient attention to those linkages. The report also identified that lending to issues of urban air quality, industrial pollution control and urban health and environmental management generally had been low and was a particular concern in the context of rising urbanisation. More coordinated actions across sectors and in given localities were also needed to build the positive linkages between environmental protection, poverty reduction and development more broadly.

Trade and the environment

At both a global and national level, trade has become an increasingly important element in development. In recent decades, more of what is produced is sold in external markets around the world confirming the growing interconnectedness of the global economy. Between 1950 and 2000, for example, world merchandise trade (encompassing manufactures, fuels, mining and agricultural products) increased almost twenty-fold, whereas merchandise production increased only six-fold (Dicken, 2011: 18). Similarly, trade now forms a larger proportion of GDP in most countries than it did 50 years ago. The contribution of developing countries to world merchandise exports has also increased, as shown in Figure 3.13.

Evidently, trade has been responsible for a significant proportion of economic growth in recent years and measures to promote further trade expansion have been central to mainstream approaches to development, as seen in previous chapters. However, just as with the debates over globalisation as a whole, there remains substantial uncertainty regarding the relationship between trade expansion, economic growth and environmental impacts. There is concern that large parts of the developing world are not experiencing the benefits of wealth generated through trade; low-income developing countries still accommodate only 1 per cent of world exports, as seen in Figure 3.13. Low-income countries also remain heavily dependent on a few products despite some diversification: just five commodities (which differ by country) account for over 70 per cent of total exports on average from these countries (World Bank, 2010b).

The geography of global production and trade is complex (see Dicken, 2011). Whilst there have been significant shifts in the last few decades, including the rising share of world exports attributed to

Figure 3.13 *The rising contribution of developing economies in world exports*

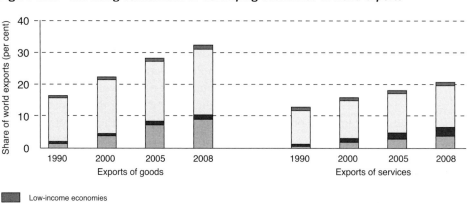

Source: World Bank (2010b).

Figure 3.14 *The uneven and concentrated geographies of production, trade and foreign investment*

....

- Approximately three-quarters of global manufacturing and services production and four-fifths of world agricultural production are concentrated in just 15 countries.
- Around one-fifth of world trade in goods, services and agriculture is accounted for by the leading two countries in each sector.
- More than 80 per cent of outward FDI stock originates from 15 countries (30 per cent being accommodated by US and UK together).
- Half of all inward FDI is concentrated in five countries (30 per cent in China and Hong Kong).

....

Source: compiled from Dicken (2011).

the developing world and the rapid economic growth of countries such as China and India, 'the geographies of production, trade and FDI remain highly uneven and strongly concentrated' (Dicken, 2011: 25), as summarised in Figure 3.14. Furthermore, there are 'persistent peripheries' within these geographies embracing 'most of the continent of Africa, parts of Asia and parts of Latin America' (p. 36).

Globally, the substantial growth in trade has been outpaced by that of Foreign Direct Investment (FDI). As identified in Chapter 1, FDI refers to investments made overseas by one firm in another for the purposes of gaining a degree of control over that firm's operations. It also includes a firm setting up a branch or subsidiary in another country. From the mid 1980s, the rate of growth of FDI started to exceed and become 'de-coupled' from the expansion of exports and trade. This is explained by the rapid growth of transnational corporations and the amount of trade that occurs between different parts of the same firm (although across national boundaries). In 2009, there were an estimated 82,000 parent company TNCs controlling around 810,000 foreign affiliates (Dicken, 2011). These TNCs account for two-thirds of world exports of goods and services and approximately one-third of world trade is transactions between different parts of same firm. This dominance of TNCs currently in world trade and the overall size of their assets (that are often larger than many countries of the world, as seen in Table 3.4) creates a number of concerns including their power to influence public policy both within international trade negotiations at the World Trade Organisation and domestically. TNCs are not accountable to the public or any elected body, but only to a small number of shareholders (Korten, 2001).

Reforming world trade became highly politicised in the 1990s as part of the wider challenge to neo-liberalism. Unprecedented protests by a wide range of development organisations, unions, NGOs and many individuals in 1999, for example, (the 'Battle for Seattle') led to the abandonment of the annual meeting in that city of the World Trade

Table 3.4 *The state and corporate power*

Country or corporation	Total GDP* or corporate sales** (millions of US$, 2008)
Belgium	471,161
Exxonmobil Corporation	459,479
Sweden	406,072
Wal-Mart stores	401,244
Saudi Arabia	375,766
BP	365,700
Argentina	307,155
Chevron Corporation	273,005
Thailand	263,772
Ford Motor Company	146,277
New Zealand	126,679
IBM	103,630
Nestle	101,466
Vietnam	97,180
Electricite De France	94,044
Deutsche Telekom	90,221
Bangladesh	89,360
Sony Corporation	76,795
Rio Tinto Plc	58,065
Luxembourg	52,296

Sources: *World Bank (2010b); **UNCTAD (2011).

Organisation (the international institution charged with setting the rules and resolving disputes in the arena of international trade). Several international NGOs (INGOs) such as Oxfam have long-standing and important campaigns concerning trade reform that focus on exposing the 'double standards' and 'rigged rules' of the international trading system that they suggest work in favour of the rich and prevent the potential for trade to reduce poverty. The nature of Oxfam's concern is illustrated in Figure 3.15 and is seen to include the activities of governments, multinational business and the World Bank as well as the WTO. There is also a recognition amongst major international institutions that the challenge of ensuring positive linkages between trade, economic growth and poverty alleviation are complex. For example, the UN Development Programme identifies that whilst trade has been an 'indispensable engine' for economic growth across the world throughout human history (UNDP, 2003: xi), the experience of trade liberalisation in recent decades suggests that 'the expansion of trade guarantees neither immediate economic growth nor long-term economic or human development' (p. 1).

There are also different views on the impact of trade liberalisation on the environment, summarised in Figure 3.16. Whilst the debate had tended to be very polarised, evidence can be identified to support both ends of the spectrum. The impact and outcomes of trade on the environment will vary, among other things, according to the nature of the product being exported and the effectiveness of domestic environmental policy. However, the operation of international trade rulings is also considered to be key. The majority of world trade takes place according to a set of rules administered by the World Trade Organisation.

Figure 3.15 The Oxfam 'Making Trade Fair' campaign

- Improving market access for poor countries and ending the cycle of subsidies, agricultural over-production and export dumping by rich countries
- Ending the use of conditions attached to IMF–World Bank programmes which force poor countries to open their markets regardless of the impact on poor people
- Creating a new international commodities institution to promote diversification and end over-supply, in order to raise prices to levels consistent with a reasonable standard of living for producers, and changing corporate practices so that companies pay fair prices
- Establishing new intellectual-property rules to ensure that poor countries are able to afford new technologies and basic medicines, and that farmers are able to save, exchange and sell seeds
- Prohibiting rules that force governments to liberalise or privatise basic services that are vital for poverty reduction
- Enhancing the quality of private-sector investment and employment standards
- Democratising the WTO to give poor countries a stronger voice
- Changing national policies on health, education, and governance so that poor people can develop their capabilities, realise their potential and participate in markets on more equitable terms

Source: Oxfam (2002).

Figure 3.16 Different views of the effect of trade liberalisation on the environment

Positive

- Trade liberalisation promotes economic growth. As societies become richer, they acquire both the will and the resources to protect the environment.
- Trade liberalisation promotes the efficient allocation of resources (including environmental resources), allowing the production of a given economic product with the least possible use of resources.
- Trade liberalisation promotes the international transfer of environmentally-preferable technologies.
- Trade liberalisation promotes the convergence of environmental standards for products and processes towards the higher levels of rich countries, and increases the markets for environmentally-preferable products.
- Trade liberalisation promotes international co-operation in other areas, notably environmental protection.

Negative

- Trade liberalisation amplifies environmental externalities through its promotion of economic growth.
- Trade often involves long-distance transport, which is one of the principal sources of environmental externalities.
- Because of competitiveness pressures, trade liberalisation will result (at best) in political drag on environmental policy making by governments, and (at worst) in an environmental 'race to the bottom' through competitive deregulation.
- Trade rules arising from trade liberalisation impede national governments in their attempts at environmental protection, either because of possible trade effects (e.g. through mandatory re-use of containers) or because of perceived discrimination (e.g. eco-labelling).
- Trade rules may inhibit the use of trade measures in multilateral environmental agreements.
- The production of some highly-traded goods (e.g. cotton, cigarettes, certain foods) is more environmentally-destructive than the production for domestic consumption which it replaces.
- Opportunities to use land for trade result in subsistence farmers being displaced onto environmentally-marginal land, where they may cause environmental damage.

Source: Ekins (2003).

The World Trade Organisation and sustainable development

The WTO was formed in 1995. It replaced the General Agreement on Tariffs and Trade (GATT) that was established in 1947 through the same statesmanship which created the United Nations, the World Bank and the International Monetary Fund. A set of international trading rules was developed for the promotion of future economic stability and development after the era of economic crisis, heavy protectionism, mass unemployment and the Depression (which had all formed part of the backdrop to the Second World War). These principles of free trade and market liberalisation continue to underpin WTO rulings.

Most countries of the world (153) are members of the WTO. Decision-making is based largely on consensus, with each member getting one vote (rather than being weighted according to economic contribution, as within the World Bank). Member states agree to two fundamental principles: of 'national treatment', under which countries must treat external participants in their economies in the same way as domestic firms; and of the 'most favoured nation', which states that any concession granted by a member to any one trading partner must be extended to all. WTO rulings are binding on its members and the WTO has the power to impose sanctions on governments through its dispute settlement system.

Although WTO rulings refer strictly to international trade policy (i.e. what happens at borders), the agreements made by the organisation have far-reaching implications for national policy-making and prospects for economic development and environmental protection within countries as well as internationally. For example, the WTO agenda has increased in scope (from the original GATT activities relating to manufactured goods) to include services (such as telecommunications and banking) and intellectual property rights and patents. It also now has authority in areas such as food safety that were formerly the preserve of national governments. Developing countries including BRIC and through groupings such as the G20 now have more influence at the WTO than before. As a result, the political importance of the WTO agenda has mounted as has the complexity of its negotiations (Reiterer, 2009). This is reflected, for example, in the length of negotiations over the WTO work programme (known as 'rounds'). It took eight years to complete the 'Uruguay Round' (1986–94) and a further seven years to agree even the shape of the latest round (agreed in Doha in 2001).

The Doha Round was to be a 'Development Agenda' (DDA). At the launch, there was substantial optimism that trade rulings could and should be made to better accommodate the development dimensions of trade and give a stronger role to developing country members in negotiations. Furthermore, there was explicit recognition within the Doha Communique that upholding an open multilateral trading system and promoting sustainable development could and must be mutually supporting. In the early stages, there was indeed some evidence of a stronger role for developing country members. At Doha, for example, trade ministers from the developing world were successful in modifying existing rules regarding trade in intellectual property rights. These are now enabling lower costs of patented medicines for treatments of HIV/AIDS, for example. At the Cancun meeting in 2003, a group of 20 developing countries left in response to the refusal of members from the EU, Japan and the US to cut agricultural subsidies. There is some sense that issues of importance to developing countries and their particular challenges have been recognised within the negotiating topics for the DDA. These include cuts being made to agricultural subsidies and in industrial tariffs on textiles and leather products that have particular significance for developing country exporters. 'Special and differential treatment' is to be considered whereby developing countries would have smaller cuts and be able to shield their agriculture and industry from full cuts. However, these were only confirmed in 2004 and talks have been substantially stalled since 2008.

Clearly, these arenas of trade have major implications for economic development in exporting countries, but also for the environment. Trade in environmental goods and services are also now part of the Doha Round, opening the prospect that current tariffs on renewable energy technologies will be subject to these rulings. As seen in Chapter 2, India, Brazil and China are now world-leading manufacturers of these technologies. The relationship between trade rulings and Multilateral Environmental Agreements are also part of the main negotiations of the WTO. Many environmental treaties place some type of restriction on international trade and could be considered strictly as violating the principles of the WTO. It was seen, for example, in Box 3.1, that under the Basel Convention, countries are obliged to prevent the import or export of hazardous wastes if there is reason to believe that wastes will not be treated in an environmentally sound manner on their destination. If a country imposes a total ban on foreign waste but in so doing subjects foreign companies to more stringent measures than may be applied to domestic companies, this could contravene the WTO principle of 'national treatment'.

Prior to the Doha Round, environmental concerns at the WTO were discussed within the Committee on Trade and the Environment (CTE). However, the CTE is not a policy-making body and its work has focused largely on the negative impacts of environmental measures on trade (i.e. environmental measures as distortions to free trade), rather than how trade liberalisation may aggravate or cause environmental degradation effects (Potter et al., 2008). For example, it is possible within WTO agreements for countries to regulate trade in certain products in order to protect human, animal or plant life or health. Any such measure must be applied to domestic as well as foreign firms (i.e. be non-discriminatory) and cannot be used as a protectionist device (i.e. must be established through scientific evidence as clearly for health or conservation ends and not for trade protection). However, WTO rulings focus on the product not the processes or methods of production. It is possible, therefore, that a country may restrict the importation of a certain good if it will cause environmental damage. What a country cannot do is stop the importation of a good which has caused environmental damage elsewhere during the course of its production: 'the way the import is produced, if it has no effect on the product as such, is not an adequate reason to discriminate against it' (Cairncross, 1995: 227). If a country wishes to impose environmental and health standards on productive activities and passes environmental legislation towards that end, that country does not have a right under the WTO articles to impose those standards on other countries. As a result, it could be argued that such countries risk making their own production uncompetitive in a world market where goods produced under less environmentally friendly conditions can still be traded.

Concerns have emerged recently in relation to 'eco' and 'carbon' labelling (discussed further below). These measures are designed explicitly to reveal to consumers the nature of the production processes underpinning goods as a way of moving towards more sustainable and lower-carbon futures. If discriminating against goods and services on the basis of the processes and methods of production (PPMs) is allowed under future WTO rulings (and possibly extended to include labour standards and even human rights conditions), there is concern that this could impact on the access of developing countries (and their generally smaller business enterprises) to world markets given their environmental standards, wage rates and working conditions inevitably reflect also their level of development (Ekins, 2003).

There are also concerns that WTO rulings continue to assume that trade takes place between countries when in fact the majority is controlled by and takes place within TNCs. Whilst TNCs are not members of the WTO and therefore have no direct power, they do have very significant economic and technical resources and make a large contribution to particular domestic economies such as the USA and within the EU. This is suggested to guarantee that their concerns will be listened to at a national level and carried forward to negotiations within the WTO (Taylor, 2003). For example, the International Chamber of Commerce is an international corporate lobby group dominated by very large TNCs including General Motors, Nestlé, Novartis and Bayer. It has permanent representation at the WTO (*New Internationalist*, 2002). In contrast, many developing countries are not able to afford the costs of maintaining diplomatic missions at the WTO base in Geneva (Oxfam, 2002). Evidently there are a number of critical challenges for the WTO in future that will significantly shape patterns and processes of development and impact on many of the world's most pressing environmental problems.

Business responses to sustainable development

As seen, many commentators are fearful of the prospects for sustainable development through further trade liberalisation and have particular concern for the dominance of large TNCs in world trade, and their power and influence within international trade policy. Fundamentally, TNCs are considered to put profit first and hold little allegiance to any particular place, community or environment. Beyond the arena of trade, TNCs have also increasingly moved into sectors relating to natural resources, energy, telecommunications, transport, water and sanitation that were formerly provided by the public sector and that are vital for poverty reduction and the prospects for sustainable development. Between 1990 and 2001, for example in the context of neo-liberalism, 132 low- and middle-income countries introduced private-sector participation in these sectors (UNCHS, 2001). However, business and the private sector are also understood to be central to meeting the global challenges of sustainable development: in providing the innovations and products for lower carbon growth, in shaping consumer tastes in future and in providing directly the finances for conservation and development projects, for example. Jonathon Porritt, a well-known environmentalist and the Founder Director of Forum for the Future that works with business as well as the public sector to promote more sustainable futures, contrasts a suggested optimism and drive amongst the most progressive business leaders with a current procrastination of political leaders and confusion and indifference of the public (Porritt, 2011). This section considers how business is changing the ways in which it works and the sources of pressure towards greater environmental and social accountability. Government policy, consumer demand and business and industry's own perception of its environmental responsibility have combined in recent years to move many companies to progress more sustainable patterns and processes of development. However, concerns remain, including issues of the power and accountability of private corporations and whether these changes represent fundamental changes in the way that business is being done, or just 'greenwashing'.

As seen in Chapter 1, public distrust (especially in US and Europe) of business and industry as a principal cause of many environmental problems emerged through the 1970s. A key response by governments was to introduce new environmental legislation, taxes and regulatory agencies to enforce, monitor and penalise processes and products that were environmentally harmful. This is the 'command and control' mechanism of environmental governance identified in Figure 3.1. These remain a key way in which business and industry are moved to more sustainable operations. However, such regulatory controls require a well-resourced and powerful regulatory infrastructure to 'police' enforcement. Attempts to impose tighter regulations in one country can also encourage industry to export its hazards elsewhere (as seen in Box 3.1). Furthermore, such policy mechanisms can be considered to be economically inefficient, to stifle competition and block innovation, and there may be little incentive for a company to invest in reducing emissions or wastes substantially below the level required by legislation. This is part of the 'ecological modernisation' case considered in Chapter 1.

A principal response of business in these early years was to appoint public relations specialists and to use the media to reshape public opinion. Around 30 per cent of the advertising budgets of many large companies in the oil, electricity and chemical industries became directed towards environmental issues through the 1980s, for example (Beder, 2002). Business also engaged in all kinds of cooperative ventures, with schools, research institutes and government assuring the public that business interests were the same as those of the environment. In the run-up to the Earth Summit, for example, 'Corporations lined up to present themselves as part of the solution, rather than the problem' (Ainger, 2002: 21). Some consider that corporations were so successful in these activities that they were able to cast doubt on the urgency of environmental problems (Beder, 2002) and even promote a public 'backlash' against the environmental movement (Cairncross, 1995).

It is considered that a turning point in the relationship between business and the environment became evident in the 1990s, as many of the larger global corporations in particular started to internalise environmental concerns as a central part of their corporate governance (Redclift, 2005). The Business Council for Sustainable Development, for example, was invited to compile recommendations on industry and sustainable development at the Earth Summit. Ainger (2002: 21) suggests that this confirmed that 'Transnational companies had made the evolutionary leap. They were no longer entities to be managed by governments, but had mutated into "valued partners" and "stakeholders" formulating global policy on their own terms.' As considered further in Chapter 6, many more businesses worldwide became concerned with issues of 'Corporate Social Responsibility' (CSR) and started reporting to their stakeholders (shareholders, employees and customers) their social and environmental performance as well as economic activities. In recent years, a number of high-profile international companies have announced wide-ranging plans that embrace commitments to reducing energy and resource use throughout their production and distribution, to suppliers through their commodity chains, to monitoring labour standards and to addressing poverty and human rights impacts of their trade, for example. In 2007, Marks and Spencer launched its £200 million 'eco-plan' termed 'Plan A'. It set out 100 commitments to achieve in five years. In 2010, these were extended to 180 specific goals across seven themes in areas of climate change, waste, natural resources, health and well-being, fair partnership, involving customers and making Plan A 'how business is conducted'. Figure 3.17 identifies some of the key commitments.

Figure 3.17 *Elements of Marks and Spencer's 'Plan A' commitments*

- Targets on energy and water efficiency by 2015 including:
  - Reduction of store energy usage by 35% per square foot
  - A 35% improvement in fuel efficiencies in deliveries to stores
  - A 25% reduction in water usage within stores, offices and warehouses
- Move towards using 50% biodiesel in lorries
- Targets to reduce, simplify and re-use packaging including:
  - Reduce home packaging by 30% by weight by 2015
  - Help customers to recycle 20 million items of clothing each year by 2015
  - Reduce the weight of non-glass packaging by 25% by 2012
- Convert produce and products to fish certified by Marine Stewardship Council, wood by Forest Stewardship Council and key clothing range to 100% Fairtrade cotton
- Launch of M&S Supplier Exchange to support suppliers by sharing best practice, stimulating innovation and helping to secure funds for investment
- Marks and Start programme of local and overseas work placements for disadvantaged groups
- Enhanced training, campaigns and product labelling to support customers and employees to make more healthy lifestyle choices

> We're doing this because it's what you want us to do. It's also the right thing to do. We're calling it Plan A because we believe it's now the only way to do business. **There is no Plan B.**
>
> (http://plana.marksandspencer.com/about, accessed 23 February 2012)

Whilst many applaud such commitments, others remain sceptical, for example about whether such commitments constitute 'greenwashing', whereby minor changes are made (and highly publicised) and the environmentally conscious consumer is exploited for further economic gain to the company. In addition, many environmental NGOs remain very concerned as to the power of large corporations to influence public thinking (and government policy) regarding the changes required for sustainable development. It is well known that corporations provide huge finances to various organisations, think-tanks, lobbyists and political front groups which are opposed to progressive climate policy and clean energy development for example (i.e. which finance 'climate sceptics/climate science denial'). Greenpeace (2010) has recently completed an investigation of Koch Industries, one of the largest private corporations in the world and dominated by petroleum and chemical interests. It has annual sales of over US$100 billion with operations in 60 countries. The Greenpeace report identifies over 40 climate denial and opposition organisations that received Koch foundation grants of over $24 million between 2005 and 2008. It also names 12 US senators who have received more than $10,000 each for federal political campaigns since 2004.

The lack of a legally binding code on international corporate behaviour is a long-standing concern particularly for INGOs. Friends of the Earth International, for example, continue to campaign for a legally binding treaty on TNCs that would ensure compliance with minimum criteria of human rights, environmental and labour standards, and place legal responsibility on the corporation for the impacts of their business practices. Although the campaign gathered momentum at the World Summit on Sustainable Development and was supported by many developing countries as well as the EU, it did not gain agreement. There are now, however, a number of not-for-profit networks that work to support the development of consistent and quality systems for reporting 'CSR' worldwide. The Global Reporting Initiative is the most widely used and includes reporting on human rights, the environment, anti-corruption, labour and all aspects of 'corporate citizenship'. It is supported by UNEP and over 1,000 companies including many leading brands such as Shell, GM and Microsoft, BMW and Coca-Cola report under the framework (Blewitt, 2008). The World Bank and the IMF (i.e. institutions beyond business) also produce sustainability reports under GRI guidelines that are continuously reviewed by a wide range of business, civil society organisations, labour and professional institutions. Similarly, over 2,500 organisations from over 60 countries including major companies like Wal-Mart are now reporting voluntarily on their greenhouse gas emissions and climate change strategies through the Carbon Disclosure Project (www.cdproject.net/, accessed 14 January 2012).

One of the most powerful tools for ensuring greater business transparency and accountability is proving to be public access to information that encourages and empowers civil society to join 'governance processes'. In recent years, various forms of labelling and certification have been developed through which industry endorses more sustainable use of resources and consumers can make more informed choices. West Germany was the first country to launch a government-sponsored environmental labelling scheme in 1978. Their 'Blue Angel' endorsement now extends to over 11,500 products in 90 categories (www.blauer-engel.de, accessed 1 February 2010) through which customers are informed of the active contribution of specific products to environmental and health protection. Similarly, in 2007, the Carbon Reduction Label was launched in the UK as a partnership between the Carbon Trust (a not-for-profit company) and the government. To join the scheme, companies have to declare the total greenhouse gas emissions for their products at each stage in their lifecycle (including production, transport and disposal). They also set commitments to reduce those emissions further over two years. If reductions are not made, the label is withdrawn. Similar schemes are being developed in Thailand, Japan and South Korea.

Other labelling schemes may be backed by private sponsors (rather than government) but typically involve similar coalitions of stakeholders in business and civil society. The idea of labelling for 'Fairtrade', for example, originated in the Netherlands in the late 1980s. It was designed to inform and attract consumers through a guaranteed label on coffee grown by small-scale farmers in Mexico (and thereby support such farmers to engage in international markets). Fairtrade Labelling International now coordinates Fairtrade labelling schemes that have extended worldwide (and the products involved expanded beyond coffee). Members include labelling initiatives in 23 countries and networks representing producers in Africa, Asia, Latin America and the Caribbean. FTLOI sets and monitors the standards required of the producers and the companies involved in the marketing and trade of FT products. It sets a minimum price that buyers of FT products have to pay to the Producer Organisations for their products that ensures a price that covers the cost of sustainable production. A further sum (the 'FT Premium') is paid by the buyer for investments in social and environmental development projects (decided upon democratically by producers and workers within the Producer Organisation). As such, the price paid by the consumer is typically higher than for non-FT produce. However, Figure 3.18 suggests that many consumers worldwide are willing to pay enhanced prices in the knowledge that such goods are less environmentally damaging and can enhance social development.

Figure 3.18 *Fairtrade on the rise*

- Global sales of Fairtrade produce by value increased 15 per cent in 2009 to an estimated total value of Euros 3.4 billion.
- An estimated 27,000 Fairtrade products are now sold in over 70 countries.
- The number of certified producer organisations are also rising, from 508 in 2005, to 827 in 2009.
- Recent surveys suggest that consumer awareness of Fairtrade products is as high as 80 per cent in many countries.

Source: compiled from FTLOI (2010).

New technologies are also supporting consumers to be more aware of the processes through which products are produced and enhance the transparency of various stages in the complex commodity chains that characterise many products (involving many stages not only in production, but in manufacture, transportation and marketing). Mobile phones are starting to be used to extract sustainability information embedded in a product's barcode and live web-based camera feeds currently provide insight into activities on the ground within Bangladeshi clothes factories supplying clothing to the Asda supermarket chain (Goodman and Wright, 2011: 16). There is also now much optimism concerning the potential of well-designed and monitored labelling schemes to enable consumers to make choices that support lower carbon emissions and in turn provide incentives for businesses at various stages in the supply chain to adopt more carbon-efficient practices. However, there are many challenges in establishing the 'green' credentials of a product in terms of quantifying carbon emissions through the production, use and disposal of a product. These are enhanced when trying to establish any suggested positive environmental and social benefits of particular products. For example, there may be a lack of data on carbon emissions at stages in the complex commodity chain, particularly in low-income countries. Furthermore at the 'ground level' of the supply chain, products (such as in forestry and agriculture) may be extracted or produced under exploitative labour conditions and in a context of a lack of political representation that are very hard to 'quantify' and indeed to change. There is also concern that extension of labelling schemes and moves towards closer carbon accounting in future may be detrimental to products from low-income countries:

> exports from low-income countries typically depend on long-distance transportation and are produced by relatively small firms and tiny farms that will find it difficult to participate in complex carbon-labelling schemes.
>
> (World Bank, 2010a: 254)

Recently, the new practice of carbon 'offsetting' has become a way that companies (and individual consumers) can demonstrate their support for reducing carbon emissions and for projects that generate environmental and social benefits, particularly within developing countries. It is a practice that depends on the recent existence of a market in carbon that is itself controversial. As seen in Box 2.6, an 'official', regulated and internationally governed market for carbon as a traded commodity was established under the Kyoto Protocol in 1997. Box 3.2 described the ways in which carbon 'offsetting' is enabled within this official market through what is termed the Clean Development Mechanism (CDM). To date, carbon offsetting has largely involved large energy-intensive industries within countries subject to such international emissions reduction targets 'offsetting' their own (high) carbon emissions through payments to support emissions reductions within other industries and countries. However, there is also an emerging 'voluntary' market in carbon. In recent years, hundreds of new companies ('offset providers') have been established through which individuals and companies make voluntary payments in relation to their own carbon emissions that are then transferred to projects, activities and technologies located elsewhere that in some way support the global reduction in carbon emissions. Many offset projects also emphasise various added social and development benefits in the host location as well as carbon reduction. This voluntary offset market (although still only approximately 1 per cent of the size of the regulated market) was estimated in 2009 to be worth US$387 million (Hamilton et al., 2010). Key companies in this field include Climate Care and the Carbon Neutral Company.

However, the mechanisms of voluntary offsetting and the social and environmental impacts of particular offset projects have been strongly contested from the outset (see *New Internationalist*, 2006 www.carbontradewatch.org and www.corpwatch.org). For example, a number of early projects were of dubious quality and impact and there was confusion amongst consumers purchasing offsets. There are ongoing debates regarding the calculations of carbon 'saved' and the price charged to the consumer to offset a particular airline flight, for example. Many projects include reforestation schemes, yet there are similar problems with calculating carbon stored within particular species and the time taken for environmental benefits to accrue. There are now new international standards being developed within the industry, promoted and overseen by the International Carbon Reduction and Offset Alliance. However, from an environmental justice perspective, carbon trading and offsetting are considered 'false solutions' (see Figure 1.14) and a new form of colonialism whereby climate policies are used to impose projects on the global South whilst northern consumers need not change their lifestyle.

International debt and the environment

Issues of debt first came to international attention in the early 1980s. In the favourable global economic climate of declining oil prices, low interest rates and buoyant world trade of the mid 1970s, heavy borrowing, largely from commercial banks, enabled the developing countries to achieve relatively high growth rates whilst still being able to service their debts. Through the late 1980s and 1990s, however, many developing countries found it increasingly difficult to service these debts in a global context of declining commodity prices, rising costs of oil and interest rate rises. The prospect of widespread defaulting on loans to commercial banks threatened the stability of financial systems internationally, prompting the involvement of the IMF and World Bank.

Whilst the burden of debt servicing for developing countries is now much lower than it was previously, as seen in Table 3.5, debt remains a significant part of the context in which sustainability has to be pursued worldwide and measures for debt relief continue to be an important policy mechanism for global economic growth. Most recently, the challenges of debt have been seen in Greece and Ireland which have had to accept emergency packages of finance and reform from the IMF. This section reviews the key implications of debt for sustainable development and identifies a number of actions towards debt reduction that explicitly seek to link debt reduction to more sustainable outcomes.

Table 3.5 *Debt service as a percentage of exports*

| Region | Debt service/exports | | |
	1995	2000	2008
East Asia and Pacific	12.7	11.4	3.9
Europe and Central Asia	10.6	18.2	18.6
Latin America and the Caribbean	25.4	38.0	14.0
Middle East and North Africa	19.7	13.3	5.3
South Asia	25.6	14.6	8.4
Sub-Saharan Africa	15.9	11.5	3.3

Source: compiled from World Bank (2010b).

Fundamentally, debt presents problems for sustainable development because monies leaving the country through interest payments are unavailable for internal, productive investments in the immediate term and 'crowd out' possibilities of other areas of public spending such as education, health and infrastructure that are essential for future economic development (Johansson, 2010). Through the 1990s, government spending on servicing debts far exceeded that on health for example, as seen in Table 3.6. The level of government austerity necessitated by debt servicing also has direct implications for the environment, reducing a government's capacity to deal with environmental protection and rehabilitation and to invest in environmental management. Furthermore, debts have to be serviced through foreign exchange, so that they can only be met by increasing exports, decreasing imports or further borrowing. Expanding exports can have very direct impacts on the environment, particularly as the principal exports for the majority of developing countries continue to be raw materials and primary commodities. The need to increase short-term productivity puts pressure on countries to overexploit their natural resources. In the long term, this 'resource mining' raises the costs of correcting the environmental destruction inflicted now and reduces the potential for sustained development in the future of those resources such as within agriculture and forestry. In an investigation of debt and deforestation through the 1980s, George (1992) established that (although the links were complex) those developing countries that deforested the fastest through the decade were, in the main, the largest debtors at that time.

Perhaps most starkly, debt raises the question of intergenerational equity: debts have to be repaid by people today in relation to loans taken on in the past by governments (and in some cases, dictators). As seen in the following sections, NGOs have been very influential in pressing for change in the way donor organisations operate and in shaping innovative programmes for more sustainable debt relief. Many continue to campaign on the moral questions of 'odious debts'

Table 3.6 *Government spending: health and debt servicing compared, selected low human development countries*

Country	Public expenditure on health (% of GDP)		Total debt service (% of GDP)	
	1990	*2000*	*1990*	*2001*
Cameroon	0.9	1.1	4.7	4.0
Pakistan	1.1	0.9	4.8	5.0
Kenya	2.4	1.8	9.3	4.1
Gambia	2.2	3.4	11.9	2.7
Nigeria	1.0	0.5	11.7	6.2
Mozambique	3.6	2.7	3.2	2.4
Sierra Leone	–	2.6	3.3	12.8

Source: UNDP (2003a).

whereby the current generations are liable for the debts taken out by corrupt regimes without any responsibility being taken by the creditors that knowingly leant to those regimes (NEF, 2008).

The first substantial attempt to *reduce* the external debt of the world's poorest and most indebted countries was initiated in 1996 by the World Bank, the IMF and a number of G8 countries. The aim of the Heavily Indebted Poor Countries (HIPC) initiative was to remove US$100 billion of the debt of the lowest-income countries. It was the first time that debts to the World Bank and the IMF were considered for reduction. The key objective was to ensure that no poor country faced an unmanageable debt burden and could achieve 'debt sustainability', defined as total external debts (to all creditors) below 150 per cent of annual exports. Before qualifying for HIPC, countries have to take part in IMF and WB economic reforms (initially to have an SAP and more recently a PRSP, for example) and engage fully with other 'traditional' debt relief mechanisms such as those available through bilateral arrangements. By 2000, only five countries had completed the qualification process and were eligible for some debt relief. Substantially in response to widespread campaigning by INGOs, particularly the Jubilee 2000 coalition, it was agreed in 1999 to 'enhance' the HIPC initiative to provide greater levels of relief, more quickly and to more countries. Significantly, many bilateral creditors, including all G8 countries, also agreed at this time to 100 per cent *cancellation* of bilateral debts owed to them. The influence of civil society groups in prompting this change was again important: in 1998, 70,000 people had formed a human chain encircling the Birmingham summit of G8 leaders to expose the unpayable and unjust nature of current debt levels. Pushing for further cancellation of bilateral and multilateral debts were also central to the Make Poverty History and 'Live 8' campaigns coalescing in particular around the 2005 meeting of the G8 finance ministers in Gleneagles, Scotland.

At Gleneagles, a new programme that allows for full, 100 per cent relief on eligible debts to the IMF and the WB was announced. To qualify for the 'Multilateral Debt Relief Initiative' (MDRI), countries must have completed the HIPC initiative process and meet the criteria shown in Figure 3.19. By the end of 2010, 32 countries had qualified and 25 of these are in Africa. Approximately US$72 billion of debt has been relieved through HIPC to date and has been particularly significant for improving the situation on the African continent. However, there remain concerns including the transparency of the

Figure 3.19 *Criteria for eligibility for the Multilateral Debt Relief Initiative*

....

● Establish a further track record (i.e. since receiving HIPC relief) of good performance under programmes supported by IMF/WB loans

● Implement key reforms agreed at the HIPC decision point

● Have implemented a PRSP for at least one year

....

qualification process for these schemes where judgements are made entirely by the IMF and WB with no participation of the debtor government or of civil society, for example. Furthermore, these mechanisms are closely linked to the performance and adoption of WB policies and prescriptions that are strongly contested, as seen above.

Debt relief for sustainable development

In the late 1980s, specific measures known as 'debt for nature swaps' (DNSs) were piloted as a way of recouping a proportion of debts and assisting developing countries explicitly to conserve the environment. Since the first projects in Ecuador and Bolivia, many countries have worked with donors, environmental groups and banks in such projects that support debt reduction and simultaneously generate additional finances for conservation (and most recently, for health). These projects take various forms, but fundamentally involve the lending agency (that may be a commercial bank or donor government) selling a portion of the debt at a discount to another donor (often an NGO), who then offers the debtor country a reprieve from that portion of debt in exchange for a commitment to a particular environmental or health project in the country.

The first ever 'debt for nature swap' took place in 1987, when Conservation International paid US$100,000 for US$650,000 of Bolivian debt (owed to commercial banks) and forgave it in return for the equivalent of US$250,000 in local currency as funds towards the Beni Biosphere Reserve (Marray, 1991). The World Wide Fund for Nature was also a pioneer of DNS and is currently working with bilateral donor governments (especially the US, but also France, Germany and the Netherlands) to swap 'sovereign' debt. A recent project between the WWF and the government of France has released US$20 million for biodiversity preservation in Madagascar.

Whilst early debt swaps were largely to establish conservation reserves they are also being used in social sectors. In 2007, the Global Fund to fight HIV/AIDS, Tuberculosis and Malaria launched its 'Debt2Health' project as a way of raising funds (beyond its traditional private donations) and in recognition that those countries facing the most severe problems in terms of health were also often those in most debt. The Debt2Health project is a three party arrangement, shown in Figure 3.20, in which creditors forgive portions of the loans owed to them by debtor countries in exchange for guarantees that 'freed up' financial resources are put into the health funds. The Global Fund then disburses those finances through their existing systems and projects. Currently, the German government is involved in an agreement with Indonesia worth Euro 50 million.

Whilst the impact of these projects on overall levels of indebtedness may be relatively small, the local environmental benefits of the funds acquired can be significant and be the basis for further economic gains such as through eco-tourism initiatives. Critics, however, point out that DNS do not change the fundamental pressures that create and perpetuate environmental degradation. Conservation benefits are long term whilst the pressure to clear forests, for example (as seen in Box 3.3) are immediate and powerful. There is concern that NGOs, multilateral banks or individual governments will not be able or willing to fund such projects to a level equivalent to the earnings which could accrue over the short term from the exploitation of those resources.

Figure 3.20 *The Debt2Health mechanism*

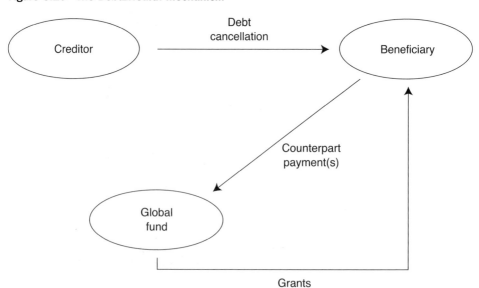

Source: adapted from The Global Fund (no date).

Plate 3.1 *Generating awareness of HIV/AIDS in Africa*

a) Zambia

Source: David Nash, University of Brighton.

b) South Africa

Source: Bjorn-Omar Evju.

National action

Governments, as seen above, are important actors in influencing cross-boundary environmental issues, in negotiating internationally to establish multilateral agreements and trade outcomes, for example, and in delivering on international aid commitments including for new climate funds. They are also responsible for establishing the policy and regulatory and institutional framework within a country and are therefore important actors in the prospects for sustainable resource management in wider arenas within their own boundaries. Governments play a key role in many mechanisms for environmental governance; in determining which uses are taxed and which are subsidised, in restricting threats posed by individual or corporate behaviour, and in allocating funds for conservation and development. Governments also set mandates for many of the regional and local agencies with responsibility for environmental protection. Through many areas of policy-making, governments also influence the kinds of choice that consumers have.

The WCED (1987) urged a key role for governments in finding solutions to environmental degradation and in ensuring that the various actors in development (including industry and consumers) behaved in the interests of environmental conservation. The Dutch National Environmental Policy Plan of 1989 is widely cited as the first attempt to convert the principles of sustainable development identified by Brundtland into concrete steps for action to change both production and consumption (WRI, 1994). Most countries worldwide now have a national report of some kind on their environment; typically national conservation strategies (NCSs) or national environmental action plans (NEAPs). The preparation of NEAPs continues to be supported by the World Bank, as identified above. The aim is to assist developing countries in moving beyond environmental reporting and the setting of specific action plans for the environment, towards integrating environmental considerations into a nation's overall economic and social development strategy. However, it is recognised that the impact of NEAPs has been uneven and many were being developed to comply with donor requirements rather than stimulating local ownership (World Bank, 2001b). A recent review suggests that many NEAPs have not been updated since being prepared and there are continued concerns regarding a lack of internal capacity to support these strategies (World Bank, 2008a).

At the Rio Earth Summit (and again 20 years later at the WSSD in Johannesburg), governments committed to the preparation and implementation of national Sustainable Development Strategies intended as national reports on activities undertaken to meet the objectives of Agenda 21. Over 100 countries are currently reporting to the UN Commission on Sustainable Development to this effect. However, as Bass and Dalal-Clayton (2004) note, early strategies tended to be all-encompassing 'perfectionist master plans', focusing on environmental dimensions rather than integrating social and economic concerns and were often remote from the realities of resource use on the ground. 'Strategy fever' is also known to have limited early reporting as developing countries in particular have been required to produce many different strategies and monitoring reports for different audiences (including to secure aid finances, as considered above), serving to undermine rather than strengthen the development of internal capacity and mechanisms for sustainable development.

Many governments worldwide are now establishing national plans for climate change and in energy, as seen in Figure 3.21. Many of these national plans are shaped by regional and international targets (such as the European Union commitments to the Kyoto Protocol) but they are also being driven by domestic development benefits such as of energy savings, reduced local air pollution and increased employment in local industries. Not only is there a key role for governments in establishing these targets, but also for providing public monies to support the necessary developments to meet those commitments, including through supporting research and providing grants, loans and tax exemptions to private investors. However, there are emerging concerns as to whether such support may constitute a form of protectionism for domestic industry and business development. For example, whilst China's commitments on renewable energy are considered to be amongst the most ambitious globally (Flavin and Gardner, 2006), one of the largest unions in the US, United Steelworkers has recently suggested that China's clean energy subsidies could contravene WTO rulings.

Government support for 'green investment' has also been adopted within many countries as part of their response to the recent economic crisis. Figure 3.22 shows the proportion of selected countries' overall 'stimulus package' (the public spending aimed to secure the necessary financial recovery) that simultaneously seeks to ensure lower carbon economic development in the future. In total

Figure 3.21 National plans on energy and climate change

Country	Climate change	Renewable energy	Energy efficiency	Transport
European Union	20 percent emission reduction from 1990 to 2020 (30 percent if other countries commit to substantial reductions); 80 percent reduction from 1990 to 2050	20 percent of primary energy mix by 2020	20 percent energy savings from the reference case by 2020	10 percent transport fuel from biofuel by 2020
United States	Emission reduction to 1990 levels by 2020; 80 percent reduction from 1990 to 2050	25 percent of electricity by 2025		Increase fuel economy standard to 35 miles a gallon by 2016
China	National Climate Change Plan and White Paper for Policies and Actions for Climate Change, a leading group on energy conservation and emission reduction established, chaired by the prime minister	15 percent of primary energy by 2020	20 percent reduction in energy intensity from 2005 to 2010	35 miles a gallon fuel economy standard already achieved; plan to be the world leader in electric vehicles; and mass construction of subways underway
India	National Action Plan on Climate Change: per capita emissions not to exceed those of developed countries, an advisory council on climate change created, chaired by the prime minister	23 gigawatts of renewable capacity by 2012	10 gigawatts of energy savings by 2012	Urban transport policy: increase investment in public transport
South Africa	Long-term mitigation scenario: emissions peak in 2020 to 2025, plateau for a decade, and then decline in absolute terms	4 percent of the power mix by 2013	12 percent energy efficiency improvement by 2015	Plan to be the world leader in electric vehicles; and expand bus rapid transit
Brazil	National plan on climate change: reducing deforestation 70 percent by 2018	10 percent of the power mix by 2030	103 terawatt hours of energy savings by 2030	World leader in ethanol production

Note: Some of the above goals represent formal commitments, while others are still under discussion.

Source: adapted from World Bank (2010a), World Development Report.

Figure 3.22 *Green stimulus spending, selected countries, 2009*

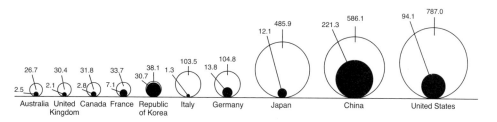

○ Size of total stimulus package ($ billions)
● Size of green share of total package ($ billions)

Source: adapted from World Bank (2010a).

approximately US$436 billion for green investments have been proposed as part of such fiscal stimuli worldwide (World Bank, 2010a) ranging from 1.3 per cent of the package in Italy to over 80 per cent in the Republic of Korea. The nature of the investments proposed differs across countries. In the Republic of Korea, a large part of the investment is for river restorations, expanding public transportation and for energy conservation in villages and schools. It is estimated that 960,000 jobs will be created. In the US, the majority are climate-change related investments including the retro-fitting of federal buildings towards greater energy efficiency and weather-proofing of homes. In total an estimated 325,000 jobs a year will be created. Clearly, the impact of these proposed investments on longer-term shifts to a lower carbon (and more economically stable) future remains unclear.

In addition to the use of subsidies to support moves towards more sustainable developments, there are a host of ways in which governments act to prevent environmentally detrimental activities, particularly through the use of legislation and taxation. These have been important (and remain widely used, as seen in Figure 3.1) in controlling the impacts of industry and agriculture on the environment through limiting particular pollutants and use of certain environmentally harmful products. Most recently, they are measures being used by governments towards promoting wider societal and cultural change and more sustainable outcomes. For example, many countries now seek to restrict vehicular use through a combination of taxation (on fuel) and legislation to support environmental goals. In France, legislation enables the authorities in Paris to restrict car use in the city when pollution levels reach certain thresholds, allowing only cars with even-numbered registration plates to be driven into the city when pollution levels merit such action (and odd-numbered ones on the next occasion). Spain has recently reduced the motorway speed limit by 10 km per hour as a means to cut petrol consumption, lower carbon emissions and achieve national commitments. London has a system of 'congestion charging' where drivers of private vehicles are charged to enter particular zones within the city. A proportion of the monies raised through this scheme are then invested to support programmes of public and environmental benefit such as cycling schemes and public transport improvements. Similar schemes using electronic toll cordons around city centres are now used in cities worldwide including in Chile, India, Norway, Singapore and Australia. Whilst there is some opposition to government interventions via legislation in individual consumption decisions, many countries including Rwanda, Ireland, Somalia, China and France now have legislation that prevents supermarkets from giving free plastic bags to shoppers because of the environmental damage these cause (such as oil used in production, time taken to biodegrade and impact on wildlife and environmental aesthetics due to littering). There are also concerns as to whether these mechanisms to promote behaviour change will lead to the kinds of change in values and attitudes that are considered key to more sustainable societies, as considered in Box 3.4.

An overarching role for governments that is understood as critical across many aspects of sustainable development is in ensuring political stability within national borders and the ability of citizens to participate in political and decision-making processes. The notion of 'good governance' now informs many arenas of development practice, including the conditions required of countries prior to lending decisions of multi and bilateral donors. Figure 3.23 summarises the components of 'good governance' as identified by the World Bank. These include characteristics of government as well as other elements of the state such as the civil services, police and courts. The thinking is that the outcomes in development depend not only macro-economic stability and good policies, but also the structures through which policies are developed and applied. Good governance as forwarded by the WB seeks to support change whereby governments are more responsive to the needs of the poorest groups, take action to control corruption and ensure political stability. It encompasses sound administration, responsiveness and accountability of all public institutions to the needs of their electorate, the access of citizens to decision-making and the

---

Box 3.4

Questioning the role of fiscal incentives for sustainable change

Fiscal incentives are being widely used as a means for changing environmental behaviour. In the UK, measures such as road pricing and congestion charges have been successful in reducing car use into cities including London and Durham. In the Republic of Ireland, the Plastic Bag Environmental Levy has led to a 90 per cent reduction in use of plastic bags since 2002. Evidently, these kinds of financial incentive have a role to play in changing behaviour, but there is also a risk of over-reliance by governments on such measures. A concern is that people may respond to the immediate signal (financial penalty) without being swayed by the underlying rationale (of unsustainable development and attitudes to pollution for example). Moreover, if people remain uncommitted to the underlying ideas, they will seek to avoid paying charges. This is confirmed by the industry that has grown up around making number plates illegible to cameras as cars enter the Congestion Zone in London and examples of illegal dumping of rubbish to avoid landfill and rubbish taxes. Further research is needed into whether and how programmes focused on changes in behaviour impact on people's attitudes and values. Are people just passively reacting to the market incentive? Could the incentives be removed and behaviours remain the same? Have values changed or have behaviours become habitualised? It is argued that this fiscal approach to change focused on incremental, sector by sector change, will not deliver the widespread changes in behaviour that are needed for more sustainable development in that they are too limited in extent, but also that they are inconsistent with what is known about environmental challenges: that they are connected and interrelated. Andy Dobson is a key proponent of a different approach to change, one through 'ecological citizenship' where behaviour is informed by deeper levels of attitude change and systemic understanding of the problems with a particular focus on justice. He suggests that behaviour driven by environmental citizenship is more likely to be sustained, but also cautions that initiatives may take longer to get going.

Source: compiled from Dobson (2009).

---

autonomy of associations of civil society including freedom of the press and the presence of democratic politics. However, it is an agenda that is widely debated and contested, among other things for how such requirements within donor lending may undermine national sovereignty and how they prioritise a western model of political democracy (see Williams et al., 2009).

Figure 3.23 *The World Bank dimensions of good governance*

- **Voice and accountability**: the extent to which citizens within a country are able to participate in selecting their government, freedom of expression and association, and free media
- **Political stability and absence of violence**: perceptions of the likelihood that government will be destabilised by unconstitutional or violent means including terrorism
- **Government effectiveness**: the quality of policy formulation and implementation and the quality of public services, the civil service and inpendendence of these from political pressure
- **Regulatory quality**: the ability of government to develop and implement policies and regulations that promote private sector development for example
- **Rule of law**: the extent to which all agencies have confidence in and abide by rules of society including the quality of contract enforcement, the police and courts and the likelihood of crime and violence
- **Control of corruption**: the extent to which public power is exercised for private gain, ranging from 'petty' to larger forms of corruption and including the 'capture' of state and government by elites/private interests

Source: compiled from Williams et al. (2009).

These few examples illustrate that the role of governments in environmental action and sustainable development is both wide and challenging. As seen in Chapter 1, a central element in neo-liberal development thinking was to 'roll back' and reduce the role of governments and the state within the economy and society. In recent years and in response to the global economic crisis in particular, there is evidence of a stronger role of governments in addressing the crisis, in protecting the public good and in capturing the ('win-win') gains for economy and environment in the future.

Non-governmental organisations and sustainable development

NGOs are highly diverse organisations, engaged in various activities and operating at a variety of levels.

> The term non-governmental organisation encompasses all organisations that are neither governmental nor for profit. What is left is a residual category that includes a vast array of organisations, many of which may have little in common. They can be large or small, secular or religious, donors or recipients of grants. Some are designed only to serve their own members; others serve those who need help. Some are concerned only with local issues; others work at the national level, and still others are regional or international in scope.
>
> (WRI, 1992: 216)

Plate 3.2 *Pressures for good governance*

a) Botswana
Source: author.

b) Cape Town, South Africa
Source: author.

NGOs are one aspect of 'civil society' that has been receiving substantial interest in the thinking and practice of sustainable development in recent years (Edwards and Gaventa, 2001). Commonly, civil society is identified as 'an arena for association and action that is distinct and independent from the state and the market, a voluntary, self-regulating, "third sector" in which citizens come together to advance their common interests' (Potter et al., 2008: 317). This broader arena of civil society also encompasses 'social movements', the term generally used to refer to coalitions and networks of actors (some of whom may be members of more formalised NGOs), which have been seen in previous chapters and sections to be important in mobilising action transnationally and promoting change around a number of inter-related issues of environmental and social justice, among other things around climate change and debt. Some authors suggest a distinction between social movements and NGOs on the basis that the former tend to work outside existing structures and work to present a more radical challenge to those than is the case with the latter (Ford, 1999).

Perhaps some of the best-known NGOs are international. Organisations like Greenpeace, the World Development Movement, Oxfam and the World Wide Fund for Nature are relatively long established, have paid professionals and tackle issues of global concern through lobbying, campaigning, direct action and the implementation of aid projects. Typically, these INGOs work for public benefit rather than for that of their members. At the other end of the NGO spectrum are many more numerous, local, 'grassroots' or 'community' organisations. Particularly in the developing world, groups of people come together for all kinds of reason to help themselves collectively: by pooling labour, to assist in gaining credit or to enable them to purchase goods in bulk, for example. People may also form community groups in response to the failure of government to provide services such as water or sewerage to low-income housing developments or in response to the unacceptability of what governments do, such as in reaction to political repression or police brutality. Quantifying such groups is difficult as many are very fluid in nature and often lack formal registration. In 2001, it was estimated that there were over 200,000 grassroots or 'community benefit organisations' in Asia, Africa and Latin America (Thomas and Allen, 2001).

National-level NGOs have regularly been formed to coordinate the activities of local organisations. In the Philippines, for example, there are many umbrella groups which seek to service and support grassroots membership (people's) organisations in particular areas of activity such as health or land reform. Other national NGOs have their origins in grassroots community initiatives, which subsequently spread to form national movements. The Green Belt Movement in Kenya, for example, started in 1977 with a single tree nursery at a primary school and now has mini green belts and community nurseries throughout the country. Its founder (the long-time campaigner for the environment, civil and women's rights and *opponent* of the government), the late Wangari Maathai, was subsequently elected to parliament in the first democratic elections in Kenya in 2002. In 2004, she received the Nobel Peace Prize for this work.

However, there have been many changes in the nature and activities of NGOs in recent years. These have blurred many of the former distinctions of NGOs such as size, whether they are working for members or for public benefit and as donors or recipients of aid. A big change in the last 20 years has been the amount of official development assistance that goes through NGOs rather than governments, as considered above. About one-third of all ODA now goes through NGOs, either provided by official donors for NGOs to use directly or into donor projects and programmes that are then run and managed by NGOs (Riddell, 2007). Further change has come as many governments worldwide have contracted NGOs to deliver services on their behalf. Many new NGOs have formed in response to the increased availability of such funds, working in collaboration with governments and donors as many governments' capacity in service delivery shrank with debt and under SAPs, for example. In this way, the proliferation of NGOs can be considered in part the outcome of the neo-liberalisation of governance through the 1990s, that is, as community participation in many instances may be being sought less for the political objectives of empowering the 'voiceless' and more for pragmatic reasons such as not relying on the state to provide (Bebbington, 2004). Most recently, there are ideas that civil society can provide public outcomes that governments cannot, such as in the UK coalition government's notion of the Big Society. In developing countries, rising affluence and the growth of the middle classes and the advent of more democratic regimes (seen starkly in countries of North Africa in the past year) have all been factors in the rising numbers and significance of NGOs. New ICT developments such as the internet and mobile phones have also enabled new alliances between NGOs and with individuals and the rapid dissemination of information and mobilisation of support.

In short, many NGOs in developing countries are now both recipients and donors of aid and are increasingly working together and with other kinds of organisation including governments and international institutions. Figure 3.24 identifies the multiple ways in which different civil society organisations now work with the World Bank to shape both aid agendas and the outcomes of particular projects. This section considers how the pursuit of sustainable development has been a key factor in the growth and importance of NGO activity and these enhanced interactions and also identifies the continued challenges and opportunities encompassed.

NGOs are not able to sign treaties, pass legislation or set targets for emissions as governments are. Historically, however, they have been a strong force in lobbying for such actions to be taken, in modifying governmental activities and in contesting the operation of international institutions such as the World Bank and the WTO. Whereas NGOs have not been represented within formal negotiations such as at the Earth Summit, they now have an official role in, for example, many UN conferences and at meetings of the Development Assistance Committee (that oversees aid from OECD countries).

Figure 3.24 *Types of World Bank (WB) and civil society engagement*

- Support to bring senior officers of Civil Society Organisations (CSOs) to annual meetings of WB and to engage in policy development
- CSO participation in country-level discussions including development of Country Assistance Strategies and Poverty Reduction Strategy Papers
- CSO recourse to Inspection Panel – where people affected by WB projects can raise concerns and request reviews
- Involvement in Bank-funded operational projects – in design, implementation and evaluation
- Civic Engagement and Social Accountability activities – funding of projects, meetings and training activities for example that aim explicitly to further strengthen civil society participation in development generally (i.e. beyond just WB projects). In particular towards rising grassroots demand for better local accountability and governance
- Grant funding to CSOs – typically small grants to youth and faith groups and community-based organisations
- Engagement with specific groups such as youth, persons with disabilities and labour unions towards mainstreaming joint working into the future
- World Bank staff attending meetings of international civil society networks such as the World Social Forum
- Collaboration around specific themes – such as water privatisation, extending urban supply and sanitation to the poor and on odious debts.

Source: compiled from World Bank (2009a).

Plate 3.3 *NGO–state collaboration in slum upgrading, Delhi, India*

Source: Hamish Main, Staffordshire University.

In 1990, only 20 per cent of World Bank projects involved some kind of collaboration with civil society organisations (CSOs). In 2009, this figure was over 80 per cent (World Bank, 2009a). However, there are long-standing questions concerning the collaboration between the World bank and CSOs: whether CSOs are merely consulted at the initial stages of a project or engaged in sustained partnerships, for example. Previous evidence suggested that the participation of CSOs is often limited to the implementation stages of projects and is therefore according to pre-identified agendas of the World Bank, with little shared control over development processes (Malena, 2000). However, the WB now in their evaluation and reporting considers not only the stage and how CSOs are engaging, but also whether there are institutional mechanisms in place such as steering committees and consultative bodies to ensure collaboration is more than a one-off engagement. Furthermore, there is now greater CSO involvement in policy dialogue with the WB, as seen in Figure 3.24.

NGOs are considered to have a number of characteristics that are thought to make them particularly suited to effecting sustainable development, including their size, their tradition of working closely with local people and their environment, and their flexibility. As seen in Chapter 1, 'participatory development' emerged as a discourse within development thinking in the early 1980s. 'Empowering' community and grassroots organisations is now promoted by various agencies as the route to an alternative development which may be more sustainable, as well as more democratic and efficient, than previous patterns and processes (Potter et al., 2008). In the following chapters, illustrations will be given of the specific actions of NGOs in rural and urban environments. The WCED recognised the key role which NGOs could have in fostering sustainable development based on their proven ability to secure popular participation in decision-making and experience is suggesting that this can best be secured through processes of planning and action which put people's priorities first rather than those defined by outside actors and agencies (Chambers, 1983). Similarly, NGOs have traditionally shown greater flexibility and adaptability than larger and more bureaucratic government institutions and experience of sustainable resource management suggests that a local body through which people's own values and needs can be discussed, planned for and acted upon is essential. Empowering the poorest sectors of society to become agents of their own development is a critical requirement of sustainability and NGOs have often worked with some of the poorest groups at the grassroots levels in areas of welfare and relief in particular. Many have subsequently moved into actions to address the underlying causes of deprivation and vulnerability, to promote self-reliant development and to facilitate development on behalf of other organisations (Korten, 1990) confirming the potential for NGOs to foster more sustainable processes (as well as patterns) of development.

A challenge for NGOs themselves may be to maintain their accountability to local communities and 'watchdog' functions as more official aid is now channelled through them and they are increasingly required to operate in commercial markets in the delivery of services. It may be that the 'better organized or more acceptable' (Mohan, 2002: 48) capture resources, for example. There is the prospect that the original missions of NGOs and power to control their own agendas may become compromised. In competing in the marketplace, smaller NGOs may be squeezed out, and NGOs generally may be forced to become more bureaucratic, leading to more finances being spent in those arenas rather than on their beneficiaries. As donors increasingly work through NGOs rather than governments, this may also dampen wider processes of democratisation in those countries as the pressure for governments to become more accountable and transparent is decreased. Critically, it cannot be assumed that all interests are served through 'community development efforts' (Potter et al., 2008); societies are highly differentiated including by gender, ethnicity and class, for example, which ensures that NGOs do not offer a simple panacea for sustainable development.

Conclusion

Many of the actions taken at various levels to promote sustainable development, as highlighted through this chapter, give cause for optimism. It is evident that many institutions of development are transforming what they do in operation, modifying the ways in which they work with other organisations and changing their internal structures. Furthermore, new institutions are being created at all levels towards the elimination of poverty and the revitalisation of economic growth, with greater weight given to environmental concerns, not least through public pressure for more information, transparency and accountability. It is also clear, however, that further changes throughout the hierarchy of institutions are required. Indeed, the capacity to generate sustainable development interventions at any level very often depends on actions at other levels. For example, community organisations require a national framework which allows local democratic processes to develop and local voices and needs to be articulated and acted upon. The effectiveness of international institutions is constrained fundamentally by national governments being willing to look beyond narrow national interests to the collective objectives that cannot be accomplished individually and by the strength of their commitments to develop rules of behaviour and to ensure compliance where necessary within their borders. Whilst significant partnerships have certainly developed in recent years, there are substantial concerns, for example, as to how the traditionally strong characteristics of NGOs in sustainable development may be compromised through new relationships with donors, governments and international institutions. There are also concerns as to how genuine and sustained the commitments of large business interests are to environmental and social responsibility. The following Chapters 4 and 5 consider how these changes in the operation and practices of core institutions in development have shaped outcomes for people and environments within rural and urban contexts of the developing world.

Summary

The notion of environmental governance is useful for understanding the complex networks of organisations and mechanisms that now influence environmental actions and outcomes.

'State-centred' international actions on the environment, trade, climate financing and debt relief remain central to actions on sustainable development but are becoming increasingly difficult to negotiate.

New, innovative networks of aid donors and means for disbursing aid have been developed in recent years. However there are challenges of coordination and coherence and concerns as to how capacity within developing countries could be undermined rather than supported.

The World Bank has been a target for environmental critics, but there have also been many changes within the organisation and in its activities towards promoting more sustainable development.

Developing economies have a rising share of world exports and a stronger role in world trade negotiations, but the geographies of production and trade remain uneven and concentrated in particular regions and countries. World trade increasingly takes place through and within transnational corporations rather than countries.

Reforming world trade to better address poverty and the needs of low-income countries remains a central campaign for international NGOs, particularly as the Doha Development Agenda at the WTO has stalled.

The business response to sustainable development is very wide-ranging, responding to a combination of government legislation, environmental taxes and subsidies, consumer demand and voluntary codes developing within the business and industry sectors. Greater public access to information can be a powerful tool for ensuring business interests are more accountable to the environments and people that their activities impact on.

The burden of debt servicing within developing countries has declined, but debt and low national incomes remain a significant part of the context in which sustainable development has to be pursued. Adherence to programmes of economic adjustment remain part of the conditions for low-income countries to access further finances.

Civil society organisations are now significant actors in sustainable development: in influencing aid policy, as donors themselves and in working jointly with other institutions to deliver development interventions. This brings new challenges as well as opportunities.

Discussion questions

- Identify the arguments for and against the World Bank now being a greener institution.

- Why are the actions of TNCs so central to the prospects for sustainable development?

- Research in more depth one of the campaigns of an INGO such as Oxfam on reforming world trade or the Jubilee Debt Campaign on debt cancellation. What have been their successes and failures?

- What do you think the significance of carbon-labelling schemes could be in future for more sustainable development?

Further reading

Dicken, P. (2011) *Global Shift: Mapping the Changing Contours of the World Economy*, sixth edition, Sage, London. A very good source for more detailed insight into how the world economy works, the key organisations involved and patterns of geographical integration, and includes how companies are responding to issues of sustainable development.

Evans, J.P. (2012) Environmental Governance, Routledge, London. A clear introduction to the field of environmental governance, the principles of and approaches in practice. A good source for more detail regarding ideas of ecological modernisation and of market mechanisms in particular.

George, S. (1002) *The Debt Boomerang*, Pluto Press, London. A classic study of how debt is detrimental to both lenders and recipients.

Mohan, G., Brown, E., Milward, B. and Zack-Williams, A.B. (2000) *Structural Adjustment: Theory, Practice and Impacts*, Routledge, London. A useful source to consider how the policies designed to address the problems of debt regularly caused greater environmental and social hardship.

Potter, R.B., Binns, J.A., Elliott, J.A. and Smith, D. (2008) *Geographies of Development*, third edition, Pearson Education Limited, Harlow. A good overview text on issues in development that includes consideration of the major institutions in development

Riddell, R.C. (2007) *Does Foreign Aid Really Work?* Oxford University Press, Oxford. A readable and balanced assessment of the positive attributes and failings of aid by an author with substantial direct experience in the aid industry.

Websites

www.worldbank.org   Home page for the World Bank on which you can search through 'topics' for details of work on the environment, sustainable development and poverty reduction strategies, for example.

www.wto.org   Home page of World Trade Organisation through which details of the Doha Development Agenda can be found.

www.oecd.org/dac   For statistics and reports on aid flows (including to the environment) from countries of the Organisation for Economic Cooperation and Development.

www.theglobalfund.org   Home page for Global fund to Fight AIDS, Tuberculosis and Malaria with details of the Debt2Health swap initiative.

www.corpwatch.org   An organisation committed to research and investigative journalism that monitors and exposes environmental and human rights violations by corporations.

www.jubileedebtcampaign.org.uk   Site for the UK organisation campaigning for debt cancellation – provides links to similar southern-based organisations.

www.tradejusticemovement.org.uk   A coalition of NGOs campaigning to make trade more fair for southern producers and countries.

www.globalreporting.org   A not-for-profit network that supports the development of consistent and quality reporting of corporate environmental and social responsibility. Most widely used by large business and international financial institutions.

4 ▸ Sustainable rural livelihoods

Learning outcomes

At the end of this chapter you should be able to:

- Understand the significance of rural development to the achievement of the Millennium Development Goals
- Understand the range of objectives that agriculture is expected to deliver for global sustainable development
- Appreciate the many interconnections between rural and urban areas, but also the different challenges for sustainable development within these contexts
- Be aware of the factors underlying the diversity and dynamism of rural livelihoods
- Understand the key characteristics of the globalised agri-food system
- Appreciate the different but related challenges of sustainable development of 'industrial' and 'resource-poor' agriculture
- Identify the principles of more sustainable rural development within research and development practice

Key concepts

Urban bias; livelihoods; de-agrarianisation; agri-food system; Green and Gene Revolutions; Farmer First; Participatory Learning and Action; social capital; Community Based Natural Resource Management.

Introduction

In 2008, for the first time in human history, the world's population became predominantly urban based. However, there are significant differences in the levels, patterns and processes of urbanisation worldwide. Currently, of the 5.5 billion people in developing countries, 3 billion live in rural areas (WB, 2008b) and it is projected that developing regions as a whole will remain predominantly rural until around 2020. Providing sustainable rural livelihoods for the current and future populations remains central to meeting both development and conservation needs within the developing world.

With over three-quarters of the world's poor living in rural areas (WB, 2008b), the MDG goal of halving extreme poverty and hunger will not be met without reducing rural poverty. As seen in previous chapters, livelihoods and well-being in rural areas are closely linked to environmental resources and ecological systems, but a large share of the world's most impoverished people also live in some of the most 'fragile' ecological zones in terms of slopes, soils and constraints set by aridity. Furthermore, the impacts of processes of globalisation and liberalisation have often hit the rural poor hardest (Mullen, 2008) and there has been further marginalisation of small-scale agriculture in recent decades (Vorley and Berdegue, 2001). Rural people generally have weaker purchasing power and thereby suffered more as currencies were devalued, for example. The collapse of extension systems (i.e. support structures) as governments had to cut their budgets, the removal of subsidised inputs to agriculture and the worsening terms of trade for agricultural goods all had adverse impacts on farmers in particular. Yet an estimated 2.5 billion people within developing regions depend (to varying degrees) on agriculture within their overall livelihoods – as smallholder producers and as wage labourers, migrant workers, herders, fishers and artisans, for example (WB, 2008b). As Chambers et al. (1989: xvii) state, 'For adequate and decent livelihoods that are sustainable, much depends on policies which affect agriculture'.

Worldwide, farming and food production is now part of a complex, increasingly global agri-food system and there have been unprecedented changes in how food is produced, distributed and controlled in recent years that raise many issues of sustainability discussed through this chapter. Whilst global food production has grown faster than world population growth and to unprecedented levels currently, 925 million people worldwide were undernourished in 2010, defined as having a calorie intake below minimum dietary requirements (FAO, 2010). Whilst the proportion of the world's population who are undernourished has fallen from 20 per cent in 1990 to 16 per cent currently, the MDG goal to halve the 1990 levels by 2015 is unlikely to be achieved globally (UN-DESA, 2010).

The challenges of sustainable rural development also include how agriculture is now expected to deliver a range of public as well as private functions worldwide – the 'multifunctionality' of agriculture

Figure 4.1 *The multifunctionality of agriculture*

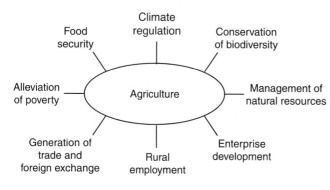

Multifunctionality refers to the interconnectedness of agriculture's different roles and functions. It recognises agriculture as a multioutput activity producing not only commodities (food, feed, fibres, agro fuels, medicinal products and ornamentals) but also non-commodity outputs such as environmental services, landscape amenities and cultural heritages.

Source: IAASTD (2009).

as shown in Figure 4.1. For example, over and above individual subsistence and food security needs, primary commodity production remains the principal source of exports and access to foreign exchange for many developing countries. Yet opportunities for agricultural expansion in many regions are limited by land and/or water shortages, as seen in Chapter 2. The loss of the world's tropical forest ecosystems (including through the expansion of agriculture) is also a global challenge in terms of climate change and for the future conservation of biodiversity (also considered in Chapter 2). In many countries and different contexts, the role of farming in the future management of natural resources and conservation is also being widely debated. In the EU, for example, as public subsidies for agricultural production are being scaled down, public concern is being expressed for the conservation of valued landscapes (and lifestyles). In short, there are a range of objectives that agriculture is expected to deliver globally, but mounting concerns over the capacity to meet these sustainably in the short and long term.

Rural and agricultural development back on agendas

Despite the evident significance of rural development in overall considerations of sustainable development, rural areas of the developing world have often been 'left out' of development theory and practice. In the 1950s/60s, most development economists focused on industrialisation as the engine of economic growth, as seen in Chapter 1. The contribution of agriculture was as a source of food and labour with an anticipated decline in the importance of agriculture in national income and employment over time. Whist rural-based strategies for development became more prominent through the 1970/80s, built on ideas of 'development from below', many development policies in fact fostered the interests of urban consumers over farmers and the rural poor. In 1977 Michael Lipton presented his now classic thesis on 'urban bias' in which he argued that the major explanation of the persistence of poverty in developing countries has been the 'anti-rural' development strategies followed. In short, his suggestion was that the urban sector benefited disproportionately from public spending in education and other services and through cheap food policies and wider public subsidies. Similarly, Robert Chambers' well-cited work of 1983 details a range of biases that have limited the understanding of the needs of rural communities and environments and compromised rural projects in practice. These include what he terms 'tarmac bias', referring to the way in which bureaucrats, academics and journalists stick to major roads and rarely venture into remote areas (thereby seeing those who are less poor and better served by the factories, irrigation, health and other services that follow main roads); 'person bias', resulting from the tendency to speak only to influential community leaders, to more 'progressive farmers' and men rather than women; and 'dry-season bias', which comes through visiting rural areas when travel is easiest (despite the wet season being the most difficult time of year for people whose livelihoods depend on cultivation).

Figure 4.2 *Declining investment in agricultural development*

- In the period 1976–78, 32 per cent of World Bank lending was to agriculture. Between 2000 and 2005, it was 6.5 per cent.
- In 1981, aid to the African agriculture sector from all bilateral and multilateral sources was $1,921 million. In 2001, it was $997 million.
- The number of technical experts employed by the World Bank in agriculture and rural development fell from 40 in 1997 to 17 in 2006.
- Public spending on agriculture is lowest in those countries where the share of agriculture in GDP is highest – comprising approximately 4 per cent of public spending.

Sources: World Bank (2008a, 2008b).

More than 30 years since these original works, it is suggested that these biases continue to be relevant (Jones and Corbridge, 2010; Chambers, 2008; Thompson et al., 2007). Figure 4.2 confirms how agricultural development was not a priority for the international financial institutions, for donors or for national governments through the latter decades of the twentieth century. Under the predominant neo-liberal approach to development, whilst countries were expected to enhance their exports of agricultural commodities, reforms of the agriculture sector per se did not feature heavily and, as identified above, served often to worsen the situation for small-scale producers and the rural poor in particular. However, there are also signs of a new interest in agricultural and rural development. In 2008, the World Bank (after a 25 year gap) made agriculture the centre of its annual *World Development Report*. It asserted that promoting agriculture was imperative to achieving the Millennium Development Goals, particularly in reducing poverty and hunger. It suggested that agriculture was 'uniquely powerful for that task' (WB 2008b: 1) through the ways that it contributes to development as an economic activity, as a livelihood and as a provider of environmental services. Similarly, donors including DFID and several African governments have produced strategy papers and new plans committing to 'pro-poor',

---

Box 4.1

The World Bank's three 'rural worlds'

In 2008, the World Bank's annual World Development Report was titled, 'Agriculture for Development'. It suggested that issues of agricultural development were imperative for and the means to achieve the MDGs and to sustaining environmental services. It sets out an agenda for how to 'use agriculture to promote development'. Key to the report was the suggestion that the way in which agriculture 'works for development' varies across countries depending on how they rely on agriculture as a source of economic growth and for poverty reduction. The report categorises countries and identifies three 'rural worlds' according to the share of agriculture in aggregate economic growth over the past 15 years and the current share of total poverty in rural areas (using the $2 a day per capita poverty indicator). On this basis, the three 'types of country' or distinct 'rural worlds' were as follows:

Agriculture-based countries – where agriculture is a major source of growth accounting for an average 32 per cent of GDP growth over the period (largely through significance of agriculture in GDP) and most of the poor (70 per cent or more) are based in rural areas. This group of countries has 417 million rural inhabitants, mainly in sub-Saharan Africa. Eighty-two per cent of the population of SSA lives in such agriculture-based countries.

Transforming countries – where agriculture is no longer a major source of economic growth, contributing on average only 7 per cent to GDP growth, but poverty remains overwhelmingly rural (82 per cent of all poor). This group (typified by China, India, Indonesia, Morocco and Romania) has more than 2.2 billion rural inhabitants. Regionally, 98 per cent of the rural population in South Asia, 96 per cent in East Asia and the Pacific, and 92 per cent in the Middle East and North Africa are in such transforming countries.

Urbanised countries – where agriculture contributes an average 5 per cent to economic growth and poverty is mostly urban. However, agribusiness and the food industry and services still account for as much as one-third of GDP and rural areas have approximately 45 per cent of the poor. Included in this group of 255 million rural inhabitants are most countries in Latin America and the Caribbean and many in Europe and Central Asia where 88 per cent of the rural populations are in 'urbanised' countries.

However, the model and the suggested policy measures that stem from it have been criticised (see Woodhouse, 2009). There are concerns, for example, about the accuracy and credibility of identifying large geographical regions within a single 'agricultural world'. Clearly, this over-simplifies the diversity that exists between particular countries. Most importantly, it diverts attention away from the complexity of rural living, the many known connections between agriculture and urban parts of the economy and the significance of non-farm incomes in rural areas, for example. The WB categorisation of agriculture-based countries suggests a rural economy comprising of smallholder farming held back by low productivity and people being 'pushed out' of a failing agriculture. In presenting three agricultural worlds, a linear model of economic growth is strongly implied – where economic development occurs in a series of clearly defined steps and the challenge is to find the technology and market incentives to push agriculture and rural society from one stage to the next.

Figure 4.3 *Agriculture-based, transforming and urbanised countries*

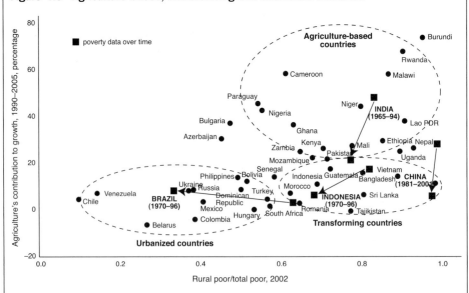

Note: Not all countries shown.

Source: compiled from World Bank (2008b).

agriculture-driven growth. However, there are some concerns as to whether the policies emerging are significantly different to the conventional 'growth focused' models for agricultural development. These raise concerns for sustainable development, as identified in Box 4.1 and considered further in the chapter.

The emphasis of this chapter is on the important shifts in research and development practice that have occurred with greater understanding of the requirements of sustainable rural development. It will be seen that overcoming the biases identified by Lipton and Chambers remain central to the prospects for sustainable rural development, in particular, those of changing the power relations between groups (within rural areas as well as between urban and rural) and fostering research and practice that empowers marginal groups.

Rural–urban linkages: complex webs of connections but persistent gaps

Importantly for sustainable rural development, there is now better understanding of the linkages between rural and urban sectors that shape both rural and urban landscapes and livelihoods (see Lynch, 2005 in this series; Tacoli, 2006). Separating people into 'rural' and 'urban' has become increasingly problematic as evidence mounts of the widespread existence of 'fluid, fragmented and multi-location households' (Lynch, 2005: 11) and of the reliance of many households (wherever located) on both rural-based and urban-based resources. Furthermore, there are sectoral interactions between rural and urban: 'historically rural' sectors such as agriculture are important in urban areas where an estimated 200 million urban residents produce approximately 15 per cent of total world food output (FAO cited in Jones and Corbridge, 2010) and urban sectors such as manufacturing are increasingly established in rural areas. As Rakodi (2002: 33) highlights, many households also live in 'peri-urban' areas 'outside the urban boundary but derive their livelihoods from work within it, while people living inside the urban boundary engage in activities such as farming, fishing, collecting wood or trading which take them to the surrounding rural areas'.

There is also now greater understanding of the importance of migration to the livelihoods of rural populations and as a key factor in the dynamics of rural economies and settlement patterns (see Lynch, 2005). Indeed, in many developing countries, there are long histories of migration such as within nomadic and semi-nomadic societies where whole households (or particular members respectively) move according to seasonal variations in resource/grazing conditions or in response to seasonal labour demands in other rural and urban contexts to return 'home' when work is complete. Other kinds of movement include 'step' migration where movement is from a village to a small town and subsequently to a larger city, and 'chain' migration where wider family members in rural areas follow their predecessors once they are established in an urban base. 'Multilocation' households are also prevalent where some household members may live in towns and others remain in rural areas.

Until quite recently, the role of migration in rural development has been seen as people responding to hardship in rural areas, that is, 'distress migration' whereby people are 'pushed' from rural areas as a response to natural calamity and other shocks. In times of drought and food shortage, for example, household members may move away to earn a wage outside rural areas and children may be sent to other villages to be cared for until food availability improves (Devereux, 1999). However, there is now more understanding of the attractions of migration as an opportunity to raise additional income, accumulate wealth and gain skills and knowledge. Consideration of 'accumulative' migration strategies has helped focus on the positive outcomes of migration including its role in stimulating rural development. In the participatory poverty assessments undertaken within the World Bank 'Voices of the Poor' analysis (see Figure 2.13), both men and women across all regions identified migration as a positive element in their lives and as an opportunity for upward economic and social mobility. This was most widely reported in Asia and Latin America and less so in Africa. There has been much interest in particular in the economic value of remittances of migrants back to their home area for individual household incomes and for wider rural development. In south-eastern Nigeria, it is estimated that the contribution of migrant associations (formed to strengthen links between migrants and their 'homes') to the construction of facilities such as schools, town halls and water points has in some cases, outstripped public investment (Bah et al., 2006).

However, understanding the motivations for migration, its impacts and the factors shaping it remains a very complex field, as summarised by Rigg (2007: 126):

> migration may be propelled by poverty, and encouraged by wealth; it
> may reflect resource scarcities at the local level, or be an outcome of
> prosperity; it may be embedded in economic transformations, or better

explained by social and cultural changes; it may narrow inequalities in source communities, or widen them; it may tighten the bonds of reciprocity between migrants and their natal households; or it may serve to loosen or break these bonds; it may help to support agricultural production; or it may be a means to break away from farming altogether.

Box 4.2 confirms that decisions regarding migration (and its economic and social value and impacts) are often made at household rather than individual level and are linked to wider processes of economic, social and cultural change that develop over time.

Box 4.2

Rural–urban migration in Lampung, Indonesia

Longitudinal research undertaken in villages in Lampung Province on the Indonesian island of Sumatra provides insights into the complexity of rural–urban linkages and their integration with processes of wider economic, social and cultural change. In the early 1990s (an era of rapid economic expansion in Indonesia) opportunities expanded for young women to work in garment and textile factories in Jakarta (located on the island of Java). By 1994, most of the (unmarried) women between 15 and 20 had left the village to work in Java. This was despite the influence of local cultural practices around family honour and prestige, which placed limits on the geographical mobility of unmarried women. As time passed, the evolution of a semi-formalised social network linking village and city allowed village-based parents to feel reassured that their daughters were 'supervised' whilst in Jakarta, and therefore not breaking local cultural codes. This was achieved largely through the establishment of a 'home town association' in Jakarta, run by senior men originally from the Lampung village, who saw their role as one of providing support for migrants in the city but who also played a role in factory recruitment, partly by attempting to subvert any sense that the prestige of village families was compromised by city migration. Initially, these young women were not expected to remit money to their family home, rather the rural-based agricultural economy provided support for them – women returning after days off spent in the village with rice and other farm produce, for example, avoiding often higher expenditures on food during their residence in the city. However, by the mid 1990s, the remitting of funds became more established and was used principally to buy gold jewellery, kitchen utensils and furniture that would help a woman 'set up home' on her return and on her (assumed) marriage, that is, it was used to improve the marriage prospects of the young woman rather than invested in any immediate needs of the rural-based household economy. However, in the late 1990s, a combination of economic, political and environmental crises complicated these remittance flows: support (in terms of food) was vital for factory workers facing unemployment and short-term contracts, whilst factory wages were necessary for shoring up agricultural incomes that were threatened by prolonged drought in Lampung.

Source: compiled from Silvey and Elmhirst (2003).

Migrancy is also just the most visible form of the many interconnecting and multidirectional flows between rural and urban. Very often, rural communities are closely tied to urban centres – both within the country and internationally – as markets for agricultural goods, sources of income, for schooling and access to many services such as credit, telecommunications, farm equipment and government services. Such linkages have intensified in recent years as incomes from farming have decreased, as non-farm rural activities have become increasingly important and through the expansion of urban centres. However, whilst rural–urban linkages exist between all rural and urban areas to some extent, the scale and strength in specific places can be very different depending on the particular economic, social and cultural transformations underway (Tacoli, 2003: Rigg, 2007). In Chapter 5, greater consideration is given to the environmental linkages that exist between rural and urban in terms of the urban needs that are supplied by rural resources and the impact of urban developments on rural areas.

Clearly, there are limitations to considering rural and urban separately given the evidently complex web of connections across space and sectors. However, there do remain important differences between rural and urban contexts that need to be understood for the development of effective approaches to sustainability. For example, the priority issues for low-income groups can be quite different: access to land, capital and labour to maintain productivity in rural contexts or environmental health and the lack of social and human capital to secure employment in an urban setting, for example. Households in urban centres also generally depend on markets to a greater extent than rural people so that security of cash incomes becomes particularly important in understanding the nature and impacts of urban poverty. The characteristics of governance can also be very different: many people in rural contexts remain physically distant and/or only very marginally associated with what are often quite new structures of local government, whereas these formal systems tend to be stronger even in smaller towns and urban centres.

There are also ways in which a continued divide between rural and urban can be identified. Figure 4.4 for example confirms the wide gap between rural and urban areas in terms of access to improved sanitation facilities. Improvements in health and education in recent years have also been faster in urban than rural areas (Jones and Corbridge, 2010). The challenge of providing these services in rural areas (that are clearly central to human well-being) are not unrelated to the problems of physical access to basic infrastructures such as roads, electricity and communications in rural areas. In turn, these factors serve also to constrain many wider opportunities for development such as through access to information, financial services, technical advice and support and markets for goods. Box 4.3 highlights the positive impacts that a more consistent supply of lighting provided by solar power can have on economic opportunity and human health for members of a weaver's cooperative in India. Yet 1.4 billion people worldwide don't have access to electricity (IEA, 2009) and many more experience limited availability and intermittent supplies.

There has been much optimism in recent years regarding the potential of mobile telephones to overcome these development constraints in rural areas. A number of opportunities of mobile phones for rural development are identified in Figure 4.5. Research to date has been largely focused on coverage and adoption rates and the impacts on the functioning of regional and national

Figure 4.4   *Disparities in urban and rural sanitation coverage*

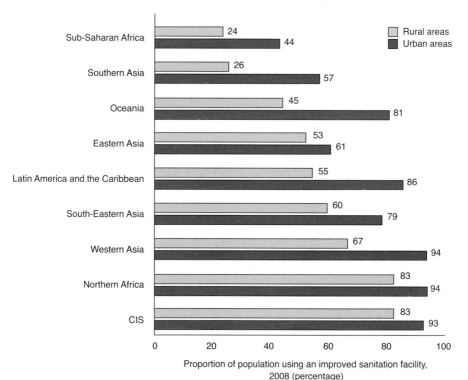

Proportion of population using an improved sanitation facility, 2008 (percentage)

Source: UN-DESA (2010).

agricultural markets. There have been fewer studies on the impacts on household economic opportunity and welfare. However, what is becoming evident is that many of the potential development benefits of mobile telephony depend on improvements to basic infrastructure in rural areas – better information on market prices are of little use if roads to markets are absent or inaccessible and if low literacy rates constrain the use of information. Fundamentally, phones still require electricity to charge them.

Box 4.3

Improving access to sustainable energy for small businesses and low-income households in India

SELCO India is a social enterprise that works with local communities, commercial banks, micro-finance institutions and donors towards finding energy solutions for under-served households and businesses in Karnataka, India. Despite the presence of major banks in the region, many are unwilling to finance solar energy projects and the majority of loans are unavailable to poorer households which do not have the required down-payments or sufficient assets to act as collateral and cannot afford the high interest rates. Several communities in the region do not have access to electricity due to their remote locations. Since 1995, SELCO has worked with banks and donors to negotiate affordable finances and to provide and maintain solar energy technologies as a sustainable alternative to often unreliable or absent rural electricity supplies. Its aim is to enhance the living and working conditions and earning capacity amongst poorer groups. It now serves over 100,000 customers.

One example is support to 102 members of a Handloom Weavers Cooperative Society in a village in southeast Karnataka. The weaving of silk sarees is a time-consuming occupation involving close concentration to deliver high-quality products. However, frequent power cuts and unavailability of light were leading to delayed deliveries and an economic loss for members. SELCO worked with the Cooperative to understand their cash flow and lighting requirements and negotiated on their behalf with banks and donors to make the required down-payments on loans to purchase solar lighting systems. Members of the cooperative now have consistent and quality lighting, are able to expand production as well as better meet deadlines whilst also meeting monthly payments to the bank.

Source: SELCO (no date).

Figure 4.5 *The potential of mobile phones for rural development*

As a communication tool:

Learn of job opportunities and market prices before travel

Improve coordination between small businesses

Ease and enhance communication between family members

Decrease costs of telephone calls

Receive/send/deposit money ('m-banking')

Foster good governance – e.g. through citizen-based monitoring ('crowd sourcing') of elections abnormalities or to report post-election unrest

Provide advanced and quicker warning of weather and pest-related shocks and decrease vulnerability

As new job opportunities:

Particularly in informal sector such as selling mobile phone credit, renting devices, selling, repairing and charging handsets

As a development tool:

Provision of mobile-enabled services ('Apps') across sectors including agriculture, health and education. Providing information and services that can:

● Monitor health outbreaks

● Extend reach of medical and agricultural workers – through providing hotlines for service users

● Send health education messages

● Improve operation of markets and address supply chain inefficiencies

● Support diffusion of innovation

Securing a livelihood in rural areas

A key feature of rural areas across developing regions is their economic, social and environmental diversity and it is important to avoid over-simplification. Indeed, many rural development initiatives have failed because they have not understood nor worked with this complexity. In 2008, the World Bank classified countries according to how they depend on agriculture as a source of economic growth and for rural poverty reduction to identify three 'rural worlds' globally, as seen in Box 4.1. Figure 4.3 confirmed that not only globally, but also within regions of the developing world, agricultural production has a greater significance in some countries than others. More importantly, within countries and in different rural settings, agriculture can play a varied role in the rural economy and for particular households. As Rigg (1997: 197) succinctly states, 'there is more to rural life than agriculture'.

Since the late 1980s, the concept of rural 'livelihoods' has been developed to better understand the many dimensions of securing a living at a household level in rural areas and how these change over time. A number of key concepts that have become central in understanding rural change are identified in Figure 4.6. The contribution of these ideas in shaping more sustainable approaches to rural development is considered in more detail below. It is now understood that livelihood diversity is very much the norm among rural households, where a range of activities that may include direct involvement in agricultural production are often combined (such as highlighted in Figure 4.7) as a means for accessing the products, labour and finance needed to securing adequate stocks and flows of food and cash to construct livelihoods. Dynamism (i.e. diversity over time) is a further fundamental characteristic of rural livelihoods. Indeed, the capacity to move the emphasis of any particular element within the livelihood system or to introduce new components has been central often to survival itself (Mortimore, 1998). Research into how poor households cope with food insecurity in times of drought, for example, has highlighted the varied adaptations to changing environmental and social circumstances which households can and do make, such as engaging in off-farm activities, including wild foods in their diet and drawing on social relationships for temporary support, such as loans of cattle.

Figure 4.6 *Key concepts in understanding rural change*

A livelihood comprises: the capabilities, assets (stores, resources, claims and access) and activities required for a means of living: a livelihood is sustainable when it can cope with and recover from stress and shocks, maintain or enhance its capabilities and assets, and provide sustainable livelihood opportunities for the next generation; and which contributes net benefits to other livelihoods at the local and global levels in the long and short term.

(Chambers and Conway, 1992: 7–8)

Rural livelihood diversification: the process by which rural households construct an increasingly diverse portfolio of activities and assets in order to survive and to improve their standard of living.

(Ellis, 2000: 15)

De-agrarianisation: the long-term process of occupational adjustment, income-earning reorientation, social identification and spatial relocation of rural dwellers away from strictly agricultural-based modes of livelihood.

(Bryceson, 2002: 726)

Figure 4.7 *Sources of rural livelihood*

- Home gardening – the exploitation of small, local micro-environments
- Common property resources – access to fuel, fodders, fauna, medicines, etc. through fishing, hunting, gathering, grazing and mining
- Processing, hawking, vending and marketing
- Share-rearing of livestock – the lending of livestock for herding in exchange for rights to some products, including offspring
- Transporting goods
- Mutual help – small loans from saving groups or borrowing from relatives and neighbours
- Contract outwork
- Casual labour or piecework
- Specialised occupations such as tailors, blacksmiths, carpenters, sex-workers
- Domestic service
- Child labour – domestic work at home in collecting fuel and fodder, herding, etc. and working away in factories, shops or other people's houses
- Craft work – basket making, carving, etc.
- Selling assets – labour, children
- Family splitting – putting children out to other families or family members
- Migration for seasonal work
- Remittances from family members employed away
- Food for work and public works relief projects
- Begging
- Theft

Source: compiled from Chambers (1997).

In recent years, there has been increased interest in the importance of non-farm incomes within household livelihood portfolios. A key issue is whether livelihood diversification operates as a 'coping mechanism' that can enhance vulnerability over time or if diversification is a source of security and sustainability of rural livelihoods and a route out of poverty. The significance of non-farm activities within the rural economy as a whole is also being assessed to consider processes of 'de-agrarianisation' of the rural sector (defined in Figure 4.6). Combined, these questions are informing work such as on adaptation to climate change (Chapter 2) and how agriculture can be made to 'work better' for economic development (World Bank, 2008b). Again, the importance of understanding differences between households and in particular rural contexts are understood as key: rural development initiatives centred on promoting non-farm activities could exacerbate rural inequality and poverty if there are barriers to poorer households such as access to finance or skills and educational constraints, or if there are simply insufficient 'low-capital entry' economic opportunities in particular areas.

Plate 4.1 *Income opportunities in rural areas outside agriculture*

a) Wage employment in brick-making, India
Source: author.

b) Desert tourism, Morocco
Source: Julie Bevis, University of Brighton.

Plate 4.1 ... *continued*

c) Fishing, Malawi

Source: Huw Taylor, University of Brighton.

There is now evidence (despite some problems of data) of a large rural non-farm economy across most countries in developing regions and an increasing role for rural non-farm activities at a household level (Davis et al., 2010). However, Table 4.1 confirms that there are substantial differences across regions (and by gender) in the relative importance (in terms of percentage of employment) of agriculture- and non-agriculture-based activities. Engagement in direct-owned agricultural production is more widespread in sub-Saharan Africa, for example, than in South Asia and Latin America where larger proportions of rural people secure a living through agriculture, but as wage labour on farms owned by others. Similarly, women's employment in agriculture is highest in SSA and relatively low in Latin America and the Caribbean. However, there are known problems of statistical evidence on gender roles in agriculture (Momsen, 2010); in many societies it may be culturally unacceptable for a woman to say that she does agricultural work, for example and indeed for the census taker to consider that she may have an economic role. Detailed fieldwork often reveals much higher levels of female participation in agriculture than identified in censuses.

Table 4.1 *Rural employment by sector of activity (percentage of adults)*

Sector of activity	Sub-Saharan Africa	South Asia	East Asia and the Pacific (excl. China)	Middle East and North Africa	Europe and Central Asia	Latin America and the Caribbean
Men						
Agriculture, self-employed	56.6	33.1	46.8	24.6	8.5	38.4
Agriculture, wage earner	4.0	21.8	9.4	9.4	10.1	20.9
Non-agriculture, self-employed	6.9	11.8	11.5	8.8	7.4	9.2
Non-agriculture, wage earner	8.6	15.4	17.4	30.9	31.3	17.2
Non-active or not reported	21.7	14.6	14.4	26.0	27.5	13.4
Women						
Agriculture, self-employed	53.5	12.7	38.4	38.6	6.9	22.8
Agriculture, wage earner	1.4	11.4	5.7	1.0	5.4	2.3
Non-agriculture, self-employed	6.8	2.9	11.3	2.8	1.6	11.7
Non-agriculture, wage earner	2.8	2.7	8.4	3.9	18.1	11.5
Non-active or not reported	32.7	64.3	35.5	53.3	46.9	51.2

Source: regional averages based on representative household surveys for 66 countries.

Research regarding the importance of non-farm activities for rural economies as a whole and for individual households has also confirmed that on-farm sources of income remain more important in rural economies in African countries compared to other regions (Davis et al., 2010). However, on-farm activities remain significant at the individual household level, with more than 80 per cent of households (in 13 of the 16 countries studied) participating in agricultural production and/or wage employment in agriculture. Wealthier households in rural areas generally have a higher level of participation and share of income from non-farm activities, although exceptions were found in Pakistan and Bangladesh where high levels of landlessness among the poor forces people into agricultural and non-agricultural wage employment.

The concept of 'agro-ecosystems' illustrated in Figure 4.8 captures aspects of both the diversity and dynamism of rural livelihood systems. The model confirms how farming is only one option for securing basic needs for food and cash in rural areas, and that farming itself may be based on a combination of livestock (including fish) and/or cropping (including forestry) systems. Further diversity stems from the limitless number of factors shaping individual farming and livelihood systems, each farming system involving the manipulation of basic ecological processes (such as the competition between species or the predation by pests), via agricultural processes of cultivation or pest control, and led by people's aspirations, priorities and goals (which may be set at the individual, community, state or international levels).

Change in the agro-ecosystem may come via the full range of environmental, economic, political and social factors across this hierarchy. For example, a change in demand for a particular handicraft item such as basketry (an economic factor) or a lack of rainfall (an environmental factor) may operate at the local level to prompt change in individual livelihoods. Box 4.3, for example, illustrated how improved access to credit/finance enabled saree makers to raise the quality of their production with economic and

Figure 4.8 *The hierarchy of agro-ecosystems*

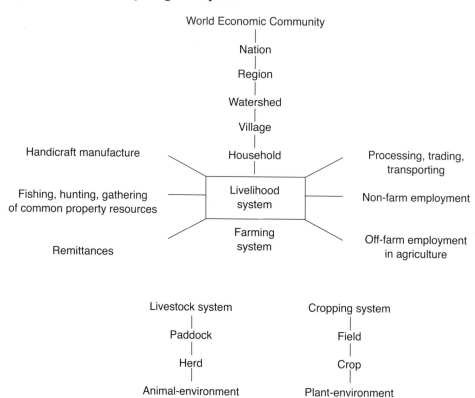

Source: adapted from Conway (1987).

social benefits for the household. Further up the hierarchy, a new national government policy concerning recommended soil conservation practices or a reduction in state capacity to finance local extension services may serve to change local cropping patterns (as may climate change).

Decisions within the (increasingly global) economic community, such as regarding agricultural subsidies, now being debated at the WTO impact on the profitability of cash-cropping decisions of farmers at the local scale. The formation of a new regional trading block between the governments of the US, Canada and Mexico (the North American Free Trade Agreement signed in 1994) led to a 50 per cent reduction in the prices paid to Mexican farmers for their corn, for example (Carlsen, 2004). As with any 'system', change in any one element has implications for the functioning of the system as a whole. The impact may be direct or indirect, large or small, immediate or delayed. Understanding rural livelihood systems within the agro-ecosystem framework, therefore, begins to illuminate the extent and nature of the challenges for sustainable rural development.

An increasingly globalised agri-food system

Few people or environments now remain outside the workings of the world economy and of the globalised agri-food system in particular. Agriculture as an economic sector and as an activity has become globalised; it is increasingly dependent on an 'economy and set of regulatory practices that are global in scope and organisation' (Knox and Marston, 1998: 337). In many places (although to varying degrees), the 'farm', which had been the core of agricultural production, is now just one part of an integrated, multilevel industrial system encompassing production, processing, marketing and distribution of food and fibre products (Redclift, 1987). Furthermore, the vertical linkages through this system have become increasingly integrated in recent years, particularly as a small number of major agri-business firms (including seed companies, food processors and supermarkets) have expanded their control over more and more stages of the system and parts of the markets within. The modern food system has been likened to an hourglass, as shown in Figure 4.9, where at the bottom are the millions of farmers and farm labourers (largely in developing regions) and at the top, the billions of consumers worldwide. In the narrow stem between these are a

Figure 4.9 *The modern food system as an hourglass*

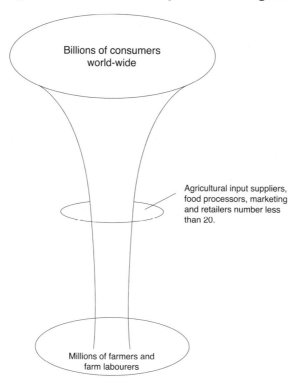

Billions of consumers world-wide

Agricultural input suppliers, food processors, marketing and retailers number less than 20.

Millions of farmers and farm labourers

very small number of major MNCs (earning a profit through all the transactions involved).

The globalisation of agriculture has had wide-ranging implications for the relationship between people and between people and the environment in rural areas of the developing world. For many commentators, the overall impacts have been generally negative, as summarised in Figure 4.10. A key concern is the degree (and increasing concentration) of control that major agri-business companies now have over decision-making throughout the agri-food system. For example, the chemical pesticide industry increasingly bought up biotechnology, plant breeding and seed interests across the world through the 1990s. In 2000, 10 corporations supplied 33 per cent of the global seed market compared to thousands of companies 20 years previously (Actionaid, 2003). Monsanto alone bought 60 per cent of the Brazilian maize seed market in the two years 1997 to 1999. The production of many agricultural commodities now involves entering into advanced contracts of some kind (including

Plate 4.2 *Cash crops for export*

a) Large-scale tea production, Indonesia
Source: author.

b) Tobacco production, Zimbabwe
Source: author.

Figure 4.10 *The negative impacts of the integration of rural producers in the global South into the global agri-food system*

- The options and capacity of governments in the global South to influence national agricultural development become increasingly circumscribed.
- National self-sufficiency in basic food stuffs has been compromised as production becomes focused on international demand for non-traditional crops.
- Farmers are tied into unequal and dependent relationships with agri-business, often with negative social and environmental effects.
- Vulnerable groups, whether farmers or workers, find their livelihoods under threat 'as they are sandwiched between downstream suppliers (providing for example, patented genetic material, chemical inputs and advice) and upstream buyers (dictating size, appearance and presentation of the product)' (p.100).

Source: compiled from Rigg (2007).

with these MNCs) regarding the types of crop grown, price and quality, for example, that introduces new risks for farmers and generally decreases their control over livelihood and natural resource management decisions. Whereas in the past, for example, farmers were dealing with several different (and smaller) firms as well as with public bodies such as government controlled marketing boards to purchase seeds and fertilisers or to sell produce, their exposure to market forces is now very different (Thompson et al., 2007). Such transactions are also now made in the context of decreasing public support for training, extension and innovation (as well as for price support for inputs and products) that has occurred through liberalisation policies. Box 4.4 confirms the very serious financial and emotional distress that farmers and their families in India are experiencing as a result of the opening of Indian agriculture to the global market in recent decades.

Different but related challenges for sustainable agricultural development

In 1987, the WCED identified three distinct types of agriculture worldwide, effectively categorised according to the degree of incorporation into the globalising world economy. Firstly, 'industrial' agriculture was identified as largely confined to the more advanced economies, but also as occurring in specialist enclaves of the developing world. This type of agriculture was identified as highly productive but dependent on heavy external inputs such as synthetic fertilisers and chemical insecticides. Multinational corporations, as identified above, had a lead role in this kind of agriculture and its expansion, dominating the supply of the key inputs, but also through establishing networks with agricultural research institutions, training colleges, and government ministries and regulatory bodies. Sustainability concerns raised at the time, such as the impacts of

Box 4.4

Farmer suicides in India

Since 1995, a quarter of a million Indian farmers have committed suicide. In 2009 alone, 17,638 farmers took their own lives – equivalent to one death every 30 minutes. Furthermore, these are official statistics that may well be an under-estimation and certainly do not account for the very large numbers of surviving family members who have been affected by those deaths. There are complex causes of suicide and factors explaining these deaths are varied in different regions. However, indebtedness is a major cause and the majority of suicides are farmers growing Genetically Modified (GM) cotton.

Since the mid 1990s, India has pursued policies of economic liberalisation that have reduced the subsidies formerly available to farmers, and have increased the presence of multinational corporations in agricultural markets generally and in GM cotton production in particular. Many farmers are now subject to increased costs and are experiencing reduced profits (and yields in some cases). There is an increased vulnerability of smallholder farmers to global market conditions (for the prices they receive) and to MNCs who control the costs and availability of many inputs to production and the costs of repayments on loans. Many farmers are trapped in a cycle of debt that creates substantial emotional distress. The Centre for Human Rights and Global Justice argues that this is a human rights crisis that has built up over the last two decades and is a direct result of the policies of the Indian government, which is legally obliged under UN human rights treaties to address this crisis.

Source: compiled from Centre for Human Rights and Global Justice (2011).

monoculture production on genetic diversity and of chemical fertilisers and pesticides on water quality, have mounted and widened in recent years. For example, the global environmental impacts of such 'high external input' systems now extend to the eutrophication of freshwater systems and hypoxia in coastal marine ecologies as a result of the application of inorganic fertilisers and pesticides. The moral issue of the existence of food surpluses being created under these systems of industrialised agriculture and the heavy state subsidies that continue to support farmers in the EU and the US, for example, are the basis of campaigns of international NGOs. Food scares such as those surrounding the link between human variant CJD and BSE in beef cattle and the health impacts of dioxins, pesticides and antibiotics emerging in the food chain are now powerful influences on consumer choice in the UK, the USA and many countries in Europe. The growth of organic food sales in major supermarkets and the rising popularity of local 'farmers' markets' (as well as the growth of Fairtrade products seen in Figure 3.18) suggest that consumers are choosing to make purchases on the basis of what they know of the production methods and commodity chains involved, that is, rejecting the environmental and health impacts of industrialised agriculture and making choices that explicitly enhance the social and economic benefits to the producer.

The second category of world agriculture identified by the WCED was 'Green Revolution' agriculture encompassing the activities of approximately 2.5 billion people in 1987 in countries of the developing world, but most widely in Asia. In areas of reliable rainfall or irrigation technologies and close to markets and sources of inputs, high-yielding varieties (HYVs) of wheat and rice in particular (and the associated packages of fertilisers and pesticides required for their production) had transformed pre-existing agricultural systems through the 1960s and 1970s. In India, for example, these technologies were central in establishing national self-sufficiency in food grains by the late 1970s. At this time, there was substantial optimism concerning the further incorporation of agriculture in developing regions into the world economy and its 'modernisation' based on moving to these 'high external input' systems of production, that is, modelling the general characteristics of industrial agriculture.

However, the 'euphoria' of the early stages of the Green Revolution (see Atkins and Bowler, 2001) faded through the 1970s as the limitations of the 'package' beyond those 'leading innovative regions' became clear. These included evidence of widening socioeconomic polarisation and negative environmental impacts in areas where such technologies were adopted (for further detail of the uneven and often gendered impacts see Potter et al., 2008). There was also a recognition that the crop types and water demands of the technologies were generally unsuited for the environmental conditions and livelihood priorities of many countries and farmers on the African continent, for example. Furthermore, evidence mounted from the agricultural research institutes that previous yield increases under traditional plant breeding methods were slowing, and problems such as chemical toxicity and changing soil carbon–nitrogen ratios were necessitating further external inputs in order to maintain yields (Pretty, 1995).

These kinds of pattern and evidence led to a call for a new 'Doubly Green Revolution'; one that is 'even more productive than the first Green Revolution and even more "Green" in terms of conserving natural resources and the environment' (Conway, 1997: 45). However, in the last decade, the debate has shifted from a 'Green' to a 'Gene' Revolution and the potential of genetically modified organisms (GMOs) to deliver further increases in global food production (Atkins and Bowler, 2001). Proponents of the Gene Revolution argue that biotechnology offers the means to feed an expanding population from a restricted land base with fewer environmental costs. GMOs are those where alien genetic material is introduced artificially rather than through traditional breeding or cross-breeding from one organism to another. Crop varieties are being created to require less pesticide, to be herbicide tolerant, to fix their own nitrogen, to yield in very challenging conditions and to be drought resistant, all of which can, for example, reduce overall energy requirements (Madeley, 2002). However, there are a range of concerns regarding GMOs in agriculture and food systems, summarised in Figure 4.11. Significantly, whereas it had been public monies and charitable foundations that had funded the experimentation and research centres of the Green Revolution, commercial biotechnology companies have become the major financers of agricultural research and development and are increasingly the architects of this Gene Revolution. The further corporate control over the food chain encompassed by these developments and the implications for small farmers who become tied into dependence on certified seeds and external inputs, for example, are a particular concern for many civil society organisations within the developing world (see Shiva, 2000).

Figure 4.12 shows the expansion of biotech crops globally since the first commercial planting in 1996. In 2010, the area under biotech crops exceeded 1 billion hectares, an expansion of 10 per cent over the previous year (James, 2010). The principal countries and crops are shown in Table 4.2. Over 90 per cent of farmers growing biotech crops are now small-scale farmers in developing countries. However, as Robinson (2004: 196) suggests, 'the application of GM "solutions" is regarded by many as the antithesis of a sustainable option for further agricultural development'. Opposition to GM crops is evident in the UK, for example, where public concerns continue to prevent both the trialling and production of such crops. Certainly, the characteristics of this externally and technology-led approach to 'sustainable' agricultural development are very different from those

Figure 4.11 *The concerns over GMOs*

Environmental concerns

1 Most agricultural crops have toxic ancestors and introduced genes could switch back on ancestral genes making agricultural crops toxic.

2 Genes inserted into GMOs will spread to other non-target organisms with unknown and unpredictable consequences.

3 We do not know enough about ecological interactions to be able to predict accurately the long-term consequences of the introduction of genes into the environment.

4 It is possible that development of herbicide-resistant plants could cause changes in the patterns of herbicide use in agriculture in ways that will be more environmentally damaging than existing systems.

5 It is difficult to predict what will turn into a plant or a weed. Once an organism becomes a pest it can be difficult to eradicate.

6 A gene does not necessarily control a single trait. A gene may control several different traits in a plant and the placement of genes is a very imprecise science in many cases.

7 A gene which is safe in one country and one soil type may behave differently under changed conditions. Therefore there are problems of scaling from field trial to commercial release.

8 The majority of the new GM crops require high-quality soils, high investment in machinery and increased use of chemicals. They do not solve the food needs of the world's poorest people.

9 GMOs encourage continuous cropping and thus discourage rotations, polycultures and the conservation of biodiversity.

Socioeconomic concerns

1 Genetic engineering is leading to the patenting of life forms, genetic information and indigenous knowledge of local ecology. Such commodification or privatisation of nature and knowledge is morally wrong.

2 Corporations are concentrating research and development on the most profitable elements of biotechnology rather than the applications that best promote sustainable development.

3 The control of the global food economy by fewer large corporations is leading to more genetic uniformity in rural landscapes.

4 Competition to gain markets and hence profits is resulting in companies releasing GM crops without adequate consideration of the long-term impacts on people or the ecosystem.

5 Without adequate labelling, consumers have no choice as to whether they eat food derived from GMOs.

6 There is no conclusive evidence that GMOs are superior to conventional crops. They may divert resources from exploring more appropriate, sustainable low-technology alternatives to intensive agriculture.

7 Using GMOs to increase agricultural productivity in the North may lead to reduced imports from the South. Farmers in the South may then turn to more environmentally damaging alternatives with adverse effects on biodiversity.

Source: compiled from Huckle and Martin (2001).

Figure 4.12 *The global area of biotech crops*

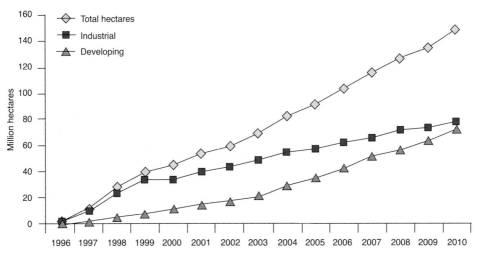

Source: James (2010).

Table 4.2 *Leading biotech growing countries and crops*

Country	Area (million hectares)	Biotech crops
USA	66.8	Maize, soybean, cotton, canola, sugarbeet, alfalfa, papaya, squash
Brazil	25.4	Soybean, maize, cotton
Argentina	22.9	Soybean, maize, cotton
India	9.4	Cotton
Canada	8.8	Canola, maize, soybean, sugarbeet
China	3.5	Cotton, papaya, poplar, tomato, sweet pepper
Paraguay	2.6	Soybean
Pakistan	2.4	Cotton
South Africa	2.2	Maize, soybean, cotton
Uruguay	1.1	Soybean, maize

Source: compiled from James (2010).

of participatory and community-based development considered below. However, Box 4.5 confirms the difficulties in practice of evaluating the sustainability impacts of GM technology.

The third type of agriculture, identified by the WCED in 1987 as 'resource poor', supported more than 2 billion people in the developing world at the time, but was less well integrated into the

Box 4.5

The sustainability of genetically modified cotton in South Africa

The Makhathini Flats area of KwaZulu Natal is one of the poorest in South Africa. Agriculture is the primary source of income for smallholder farmers cultivating plots typically of between 1 and 3 hectares and cotton is the principal cash crop. However, cotton yields were relatively poor with key constraints on production including drought, pest attacks, access to irrigation and limited options in terms of input supply and marketing. In 1998–99, resource-poor farmers started adopting Bt cotton following large commercial farmers. By 2004, most small-scale farmers in the area were growing a Bt cotton variety. A study over six seasons found that yields for Bt cotton were significantly higher and had resulted in increased income that was being used in particular to invest in children's education. However, overall costs of production of Bt and non-Bt cotton were broadly similar: for example, whilst pesticide and insecticide use declined (saving money and labour costs), there were higher labour costs involved in harvesting the greater yields.

Ongoing issues surrounding the contribution of GM cotton to agricultural sustainability include:

● Whilst use of pesticides has declined (with health, economic and environmental benefits), will Bt resistance to pests break down over a longer time period?
● Will increased cotton yields under Bt varieties continue? Will this lead to a dampening of market prices for cotton in future? Could this happen with any yield/revenue-enhancing technology (therefore not being GM-specific)?
● Bt seed is significantly more expensive than non-GM varieties and risks of crop or market failure remain.
● Adoption of Bt cotton varieties has not addressed a key sustainability issue of the 'narrowness' of the livelihood base of small-scale farmers in Makhathini, that is, the heavy reliance on cotton production per se within the household economy. This in combination with the vulnerability to drought could be considered the greatest threat to sustainability in the region. 'This, however, was the case prior to Bt's introduction and has not been caused or exacerbated by that technology' (Morse and Mannion, 2009: 243).

Source: compiled from Morse and Mannion (2009).

world economy and into the agri-food system in particular. It was in these areas that the major challenges for sustainable rural development were identified to lie; where poverty, economic and ecological 'marginality' was concentrated and which had generally been 'forgotten' through the 'biases' in development thinking and policy identified above. As Pretty (1995: 31) suggested:

> Farming systems in these areas are complex and diverse, agricultural yields are low, and rural livelihoods are often dependent on wild resources as well as agricultural produce. They are remote from markets and infrastructure; they are located on fragile or problem soils; and less likely to be visited by agricultural scientists and extension workers or studied in research institutions.

The remaining sections of this chapter focus on the challenges of sustainable rural development in these areas in more detail. However, whilst these three categories of world agriculture remain important for understanding how the challenges of sustainable agricultural development are different globally, there are now growing concerns as to whether agricultural policy and practice worldwide are able to respond to the complex and diverse challenges of the twenty-first century. In short, it is considered that the dominant approach to world agriculture has been led by concerns for further growth and the aggregate benefits of raising production whether in industrial or small-scale agriculture. In so doing, the distributional aspects of who benefits and who loses (central concerns in sustainable development) are underestimated, and moreover, opportunities for alternative routes out of poverty are being missed. In 2007, in a major review of the workings of the global agri-food system and pathways for more sustainable development, a typology of 'Five Rural Worlds' was developed (Figure 4.13). In contrast to the WCED categorisation (and that of the World Bank in Box 4.1), the typology centres on rural peoples' livelihoods and the differences between households. Whilst this research confirms the continued concentration of poverty and marginality within areas of 'resource poor agriculture', it also makes clear:

> that poverty is located unevenly across and within rural populations, that policy in and for food and agriculture affects different groups in different ways and that the actions of one rural group can improve or impair the livelihoods of others.
>
> (Thompson et al., 2007: 15)

It also confirms the importance of understanding agriculture as one part of people's livelihoods and in rural economies (as discussed above) and of the need to consider the challenges and opportunities for more sustainable rural development in particular environmental, economic and social contexts:

> Not all developing country farmers need to end up growing French beans or carnations through out-grower schemes for the European market, even if this is a good plan for some farmers in some places. There are other opportunities that need to be sought out, and the generalised models and universalised policy prescriptions do not help in the search for these pathways to sustainability, particularly in a period of rapid change and uncertainty.
>
> (Thompson et al., 2007: 31)

Figure 4.13 *Five 'Rural Worlds'*

Rural World 1 households and enterprises engaged in high-value, export-oriented agriculture, make up a very small minority of rural households and firms in the developing world. In addition to their land and other holdings, producers and firms have direct access to finance, risk management instruments, information and infrastructure necessary to remain competitive in their business operations. Most have an influential voice in national policies and institutions affecting their enterprises and close ties to buyer-driven value chains associated with global agriculture.

Rural World 2 accounts for a substantial number of rural households and agricultural firms in the developing world. They are frequently part of the local elite but have little influence at the national level. They have sizeable landholdings often devoted to both commercial and subsistence agriculture. They previously had access to basic services, such as finance, but with the advent of liberalisation and the consequent withdrawal of the state from a direct role in agriculture, the availability of these services declined rapidly.

Rural World 3 households – fisherman, pastoralists, smallholders and associated micro-enterprises – are survivalist. Food security is their main concern, and their small production units are almost totally dedicated to home consumption. Their assets are poorly developed, and they have very limited access to key services (e.g. credit) that would enable them to increase returns to their assets.

Rural World 4 households are landless or near-landless, frequently headed by women, with little access to productive resources other than their own labour. Sharecropping or working as agricultural labourers for better-off households in their communities in Rural Worlds 1, 2 and 3 is perhaps the most secure livelihood option for many of them. For others, migrating to economic centres on a daily, seasonal or even permanent basis is their best hope for survival. But their low education levels are a major barrier to migrating out of poverty.

Rural World 5 households are chronically poor. Most have sold off or been stripped of their asset holdings during periods of crisis. Remittances from relatives, community safety nets and government transfers are vital to their sustenance. As a result of the HIV/AIDS pandemic, many more households are facing this precarious situation, particularly in Africa. Entrenched gender inequalities exacerbate the problem. Social exclusion often typifies the relationship of Rural World 5 to the larger community. Social protection programmes, including cash and in-kind transfer schemes, will be critical for this group for some time.

Source: Thompson et al. (2007).

Towards sustainable rural development

The importance of the local context

Individuals ultimately make the land-use decisions upon which sustainable agricultural development depends. They do so, however, within the context of a variety of natural (environmental and ecological) and structural (including the world economy) forces, as identified in the hierarchy of agro-ecosystems above (Figure 4.8) and in Box 4.4. Through these illustrations it is seen that whilst the range of options in decision-making for particular people regarding natural resource use may depend closely on factors of finance, access to technology and political power, it can also be suggested that behaviour is not fully determined, even amongst the poorest farmers:

> Political, social and economic forces do operate; but when they are dissected, sooner or later we come to individual people who are acting, feeling and perceiving . . . all are to some degree capable of changing what they do . . . the sum of small actions makes great movements.
>
> (Chambers, 1983: 191–2)

Whilst this 'populist' stance can be criticised for underestimating the broader context of individual livelihoods, that is, the political, economic and social structures and processes that underlie land use decisions and practices leading to resource degradation (the focus of political ecologists identified in Chapter 2), there is much evidence from analyses of sustainable agricultural development that the local level is a key arena for success. For example, ecology-based alternatives to the 'industrial' agricultural model are now being developed worldwide. Whilst various terms are used including 'low external input', 'organic' and 'biodynamic' farming, they all centre on integrating ecological processes and principles into the way that food is produced, to prevent environmental harm in particular, but also in ways that are tailored to the particular social, economic and environmental priorities and circumstances at the local level. They start with the understanding that spatial variability and complex ecological dynamics are key properties of the agri-farming system (Thompson et al., 2007). Whilst the technologies and specific practices will be different in particular places, key elements include the use of locally adapted resource-conserving technologies that deliver multifunctional benefits (jointly providing food and other goods for farmers and markets but also contributing to wider public goods such as flood protection and habitat conservation); using the knowledge and skills of farmers (improving self-reliance and substituting human capital for costly external inputs); and making use of people's collective capacities to work together to solve common agricultural and natural resource problems (Pretty, 2008).

The types of resource-conserving technology which are delivering favourable changes to several components of the farming system simultaneously are highlighted in Figure 4.14. Techniques such as intercropping and agro-forestry demand a close understanding of the specifics of local ecological and environmental conditions, substantial

Figure 4.14 *Agricultural technologies with high potential sustainability*

Intercropping	The growing of two or more crops simultaneously on the same piece of land. Benefits arise because crops exploit different resources, or interact with one another. If one crop is a legume it may provide nutrients for the other. The interactions may also serve to control pests and weeds.
Rotations	The growing of two or more crops in sequence on the same piece of land. Benefits are similar to those arising from intercropping.
Agro-forestry	A form of intercropping in which annual herbaceous crops are grown interspersed with perennial trees or shrubs. The deeper-rooted trees can often exploit water and nutrients not available to the herbs. The trees may also provide shade and mulch, while the ground cover of herbs reduces weeds and prevents erosion.
Sylvo-pasture	Similar to agro-forestry, but combining trees with grassland and other fodder species on which livestock graze. The mixture of browse, grass and herbs often supports mixed livestock.
Green manuring	The growing of legumes and other plants in order to fix nitrogen and then incorporating them in the soil for the following crop. Commonly used green manures are *Sesbania* and the fern *Azolla*, which contains nitrogen-fixing, blue-green algae.
Conservation tillage	Systems of minimum tillage or no-tillage, in which the seed is placed directly in the soil with little or no preparatory cultivation. This reduces the amount of soil disturbance and so lessens run-off and loss of sediments and nutrients.
Biological control	The use of natural enemies, parasites or predators, to control pests. If the pest is exotic these enemies may be imported from the country of origin of the pest; if indigenous, various techniques are used to augment the numbers of the existing natural enemies.
Integrated pest management	The use of all appropriate techniques of controlling pests in an integrated manner that enhances rather than destroys natural controls. If pesticides are part of the programme, they are used sparingly and selectively, so as not to interfere with natural enemies.

Source: Conway (1997).

Plate 4.3 *Harnessing scarce water resources for agricultural production in Tunisia*

a) Tabia and jessour irrigation

Source: author.

b) Water control in the El Guettar oasis

Source: author.

management skills and access to information, and often significant levels of investment and cost adjustment. Coordinated actions at the local level are necessary to ensure that the sustainable development measures of one person or section of the community are not to be compromised by the actions of others.

In 1988 Chambers put forward five interdependent prerequisites for sustainable rural development on the basis of analysis of apparently successful and sustainable projects in the developing world. These principles, listed in Figure 4.15, have come to inform much rural development research and practice in areas within and beyond agriculture. The importance of the local context is confirmed, for example, in prioritising local people's issues and concerns but also through the specifics of the values and commitments of staff. However, the more detailed analysis in the following sections indicates that the conditions for sustainability extend also beyond the local level. They will include national systems of tenure which promote security and encourage long-term investments, for example and research and planning institutions that encourage and give value to community voices at all levels. Furthermore, experience is suggesting that the challenges of sustainable rural development often lie in the relation between these levels of action: in how women, for example, can be empowered to address broader strategic needs that tackle gender relations rather than 'simply' their practical, immediate needs as defined by existing gender roles (Pearson, 2001) or how local community based organisations can work with donor organisations and state institutions (Bebbington, 2004).

Figure 4.15 *Lessons for the achievement of sustainable rural livelihoods*

1 A learning-process approach
2 People's priorities first
3 Secure rights and gains
4 Sustainability through self-help
5 Staff calibre, commitment and continuity

Source: Chambers (1988).

Putting Farmers First

The basic model for agricultural development encompassed within both the Gene and Green Revolutions encapsulates what has been termed a 'blueprint' or 'transfer of technology' approach to planning: in short, new technological packages are developed in research stations and laboratories for transfer via extension systems to farmers. Over the years, criticism of this approach was important in generating the call for 'another' or 'alternative' development (as seen in Chapter 1). These gave recognition, for example, to local practices and knowledge and emphasised the questions of power and participation that are central to achieving sustainability. The 'Farmer First' movement aims to put farmers (broadly defined to include not just sedentary, smallholder farmers, but also farm workers, pastoralists, forest dwellers, fisherfolk and other small-scale producers of food and feedstuffs) at the centre of agricultural innovations and rural development through participatory research and planning processes. In direct contrast to the blueprint approach, Farmer First is based on a 'learning-process approach' where farmers and scientists collaborate, where technologies evolve and are adapted with experimentation and experience and where projects are continuously modified rather than being held to a rigid set of aims and procedures. The key features and contrasts between these approaches are identified in Figure 4.16. Farmer First approaches have had widespread implications (within and beyond agriculture and rural development) for how research is undertaken, how projects are designed and the value of particular kinds of 'expertise', for example and have been the basis of many innovative and successful developments. Most recently, there is understanding of the many roles and relationships that farmers now have, that is, beyond those with 'outside professionals' such as scientists and development practitioners, as entrepreneurs involved in business, trading and marketing or as organisers and advocates for others in the community. As such, farmers are involved in many forms of collaboration and networks of public, private and community-based organisations. For those working 'Beyond Farmer First' (Scoones and Thompson, 2009, 1994) the emphasis is now on innovation and learning in this new and rapidly changing context.

However, putting the Farmer First approach into practice has proven challenging. A key requirement, for example, is to understand and value local priorities, yet as already discussed, understanding the 'realities' of resource-poor agriculture has generally not been the focus of research and development. Figure 4.17 identifies a number of core ways in which farmers' and scientists' priorities may differ. Robert Chambers (1983, 1993, 2008) has been influential in detailing the powerful forces which tend to perpetuate 'first' ('scientist' or 'outsider') priorities, those which start with 'economies not people', with the 'view from the office not the field' and tend to lead to centralised, standard prescriptions for change, rather than the priorities of the 'last' (i.e. farmers) and the substantial 'reversals' that are required in what he terms 'normal' research and development

Figure 4.16 Changing approaches to agricultural research and development

	Transfer of Technology	Farming Systems Research	Farmer First/Farmer Participatory Research	People-centred Innovation and Learning
Era	Long history, central since 1960s	Starting in the 1970s and 1980s	From 1990s	2000s
Mental model of activities	Supply through pipeline	Learn through survey	Collaborate in research	Innovation network centred on co-development; involving multi-stakeholder processes and messy partnerships
Farmers as seen by scientists	Progressive adopters, laggards	Objects of study and sources of info	Colleagues	Partners, collaborators, entrepreneurs, innovators, organised group setting the agenda, exerting demand: 'the boss'
Scientists as seen by farmers	Not seen – only saw extension workers	Used our land; asked us questions	Friendly consumers of our time	One of many sources of ideas and information
Knowledge and disciplines	Single discipline driven (breeding)	Inter-disciplinary (plus economics)	Inter-disciplinary (more, plus farmer experts)	Extra/trans-disciplinary – holistic, multiple, culturally rooted knowledges
Farmers' roles	Learn, adopt, conform	Provide information for scientists	Diagnose, experiment, test, adapt	Empowered co-generators of knowledge and innovation; negotiators
Scope	Productivity	Input-output relationships	Farm based	Beyond the farm-gate – multi-functional agriculture, livelihood/food systems and value chains across multiple scales, from local to global; long time frames

continued . . .

Figure 4.16 ... *continued*

	Transfer of Technology	*Farming Systems Research*	*Farmer First/ Farmer Participatory Research*	*People-centred Innovation and Learning*
Core elements	Technology packages	Modified packages to overcome constraints	Joint production of knowledge	Social networks of innovators; shared learning and change; politics of demand
Drivers	Supply push from research	Scientists' need to learn about farmers' conditions and needs	Demand pull from farmers	Responsiveness to changing contexts – markets, globalisation, climate change; organised farmers, power and politics
Key changes sought	Farmer behaviour	Scientists' knowledge	Scientist–farmer relationships	Institutional, professional and personal change: opening space for innovation
Intended outcome	Technology transfer and uptake	Technology produced with better fit to farming systems	Co-evolved technology with better fit to livelihood systems	Capacities to innovate, learn and change
Institutions and politics	Technology transfer as independent: assumed away	Ignored, black boxed	Acknowledged, but sometimes naïve populism	Central dimensions of change
Sustainability	Undefined	Important	Explicit	Championed – and multi-dimensional, normative and political
Innovators	Scientists	Scientists adapt packages	Farmers and scientists together	Multiple actors – learning alliances

Source: Scoones and Thompson (eds) 2009.

Figure 4.17 *Where farmers' priorities might diverge from those of scientists*

	Priorities	
	Scientists	Resource-poor farmers
Crops	– yield – compatible with machine harvesting – single variety	– flavour – local marketability – multiple variety cropping
Cropping systems	– mono-cropping – high external input – high yield	– diverse cropping – low external input – yield less important
Management	– maximise production – maximise growth	– minimise risk – livelihood security
Use of labour	– minimise labour input	– use all family labour
Constraints	– meeting demands of scientific community – project cycles – meeting demands of donors	– meeting traditional obligations – maintaining good community relations

learning and practice. A simple illustration is that aid agency bureaucracy can lead to pressure to produce a portfolio of projects quickly and to spend budgets by deadlines, giving little time within the project process to be open to changing conditions and experience from practice or for projects to evolve. Too often, in the past, 'outsiders' have assumed that they knew what poor people wanted when, in practice, the priorities of farmers and of different groups within the local community would vary quite widely.

Reversals in 'learning' are also required, whereby the value of indigenous knowledge, skills and technologies is recognised in combination with 'conventional' science, as discussed in Box 4.6. As identified recently within the Millennium Ecosystem Assessment (MEA, 2005: 24):

> Traditional knowledge or practitioners' knowledge held by local resource managers can often be of considerable value in resource management, but it is too rarely incorporated into decision-making processes and indeed is often inappropriately dismissed.

The importance of using the knowledge and skills of farmers for more sustainable alternatives to conventional models of agriculture development are also a characteristic of the agro-ecology-based alternatives considered above and are a central element in defining the 'Doubly Green Revolution':

> Whilst the first Green Revolution took as its starting point the
> biological challenge inherent in producing new high-yielding food
> crops and then looked to determine how the benefits could reach the
> poor, this new revolution has to reverse the chain of logic, starting
> with the socio-economic demands of poor households and then seeking
> to identify appropriate research priorities. Success will not be achieved
> either by applying modern science and technology, on the one hand, or
> by implementing economic and social reform on the other, but through
> a combination of these that is innovative and imaginative.
>
> (Conway, 1997: 42)

Since the 1990s, many methods rooted in the philosophy of 'Farmer
First' have been developed to explore local issues and realities and to
give greater voice to a wider set of local interests. These include a
greater reliance on qualitative techniques (interviews rather than
surveys, for example) but also oral and visual methods rather than
written accounts. 'Participatory Learning and Action' (PLA) is now

---

Box 4.6

The value of indigenous technologies

> Rural people's knowledge and scientific knowledge are complementary in their
> strengths and weaknesses. Combined they may achieve what neither would
> alone.
>
> (Chambers, 1983: 75)

During the 1950s and 1960s, there was tremendous optimism for the role of western
science in raising agricultural production throughout the world, encapsulated in the
research and extension activities associated with the 'green revolution'. The locus of
research was the experimental station and the challenge was to transfer the new
technology to the farmers' fields. When it subsequently became clear that farmers were
unable to gain yields on their own farms comparable to those achieved at experimental
stations, the 'blame' was passed between 'ignorant farmers' and 'poor extension
services'.

It is now appreciated more widely that rarely do farmers fail, through ignorance, to effect
land-use decisions which will raise productivity or conserve resources. Rather their
behaviour is, more regularly, rational in the light of their political-economic, social and
environmental circumstances. It is now thought that research conducted at experimental
stations has limitations for solving the 'real-life' problems of the farmer (particularly the
resource-poor farmer). Scientists have an important role to play in conducting research

about a problem, for example how potatoes grow. But for *solving* a problem, such as how to grow potatoes, it is thought that farmers in fact have a lot to teach scientists (Chambers et al., 1989). The problem for research and extension activities, therefore, becomes not how to transfer technology from research station to farmer but how to close the gap between the two so that insights from both can be shared and built upon.

However, as more has become known regarding the ways in which farmers learn and experiment, often in very contrasting ways to modern science, it has also become clear that there are differences amongst rural people in terms of their knowledge and power. 'The issue is not just "whose knowledge counts?", but "who knows who has access to what knowledge?"and "who can generate new knowledge, and how?"' (Chambers, 1993: xv). Not only, therefore, are there substantial and continued challenges in instilling changes in attitudes, behaviour and methods in the work of institutions and extension agents, but new insights are required into how those who are variously excluded at the local level can be:

> strengthened in their own observations, experiments and analysis to generate and enhance their own knowledge; how they can better seek, demand, draw down, own and use information; how they can share and spread knowledge among themselves; and how they can influence formal agricultural research priorities.
>
> (Chambers, 1993: xv)

the umbrella term used to refer to the tools and ways of working that have done much to move understanding forward of the knowledge, values and priorities within local communities and the 'complex, diverse and risk-prone environments of resource-poor people' (Scoones and Thompson, 1994: 4). PLA is centred on trying to see the world from the point of view of those directly affected by development (Mohan, 2008). PLA is most commonly used in natural resource management and in agriculture, but also in programmes addressing empowerment, equity and rights issues (Chambers, 2008). Through the types of research tool and approach identified in Figure 4.18, more appropriate research priorities and more inclusive approaches to development interventions are being realised. Box 4.7 shows how PLA approaches are being used to support local capacity to adapt to future climate changes in the Andes.

However, there are also continued challenges for sustainable rural development despite the substantial progress made through these alternative approaches. There is concern, for example, that the 'frenzied levels of global interest in participatory methodologies' (Guijt and Shah, 1998: 4) through the 1990s led to the production of so many handbooks, guides and courses that a 'manual and

Figure 4.18 *The major components of participatory learning and action*

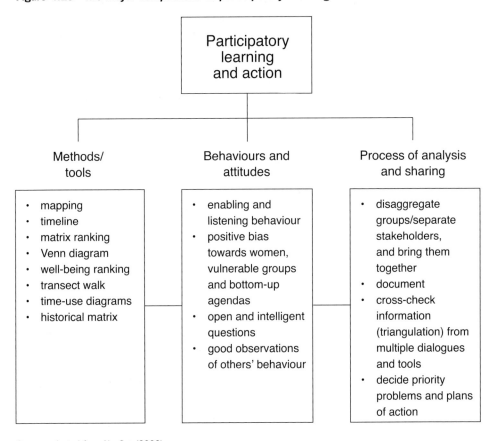

Source: adapted from Neefjes (2000).

Box 4.7

Supporting adaptations to climate change in the Andes

Successful farming in the Andes region is already a substantial challenge, requiring for example, complex soil and water management and a diversity of crops, varieties and planting schemes adapted to altitude, daily temperature extremes and seasonal weather variations. Climate change is expected to aggravate substantially what is already a precarious situation; melting of the Andean glaciers is predicted within 15 years, droughts, flooding, wind and cyclones are expected to become more common and stream flows will decrease, for example. In 2005, a project was started in two highland locations in Bolivia and Ecuador amongst the poorest regions of the Andes, combining the work of two national NGOs, the Ecuadorian Network for Community-based Natural Resource Management (MACRENA) and the Bolivian Programme for Integrated Development of Potosi (PRODINPO), and an international NGO, World Neighbours. The project adopted a participatory approach involving communities in locally led learning and practical activities, based on the farming experience, local knowledge and technologies already in use and starting in small simple ways then diversifying over time. The aim is to enhance knowledge of climate change and improve resilience and opportunities for coping with its impacts. Initial activities centred on understanding local knowledge on climate and priorities for the future. Water management was found to be key – communities were already suffering from both droughts and floods at different times of year – and was central to learning how to cope with climate change. Practical activities were designed (including visits to particularly innovative farmers, site testing of alternative types of irrigation, composting and water harvesting, for example) towards building knowledge and evaluating options. Over time, support and facilitation was increasingly to collective rather than individual activities; where participants worked together to design and install soil and water conservation efforts on different farms, to reforest a vulnerable hillside, plant windbreaks or to establish their own savings and loan funds to help finance local purchase of fencing materials, micro-irrigation and water harvesting technologies, for example. In northern Ecuador, four communities living on the Ilalo Volcano which had initially been involved in very local priorities of soil conservation and water harvesting for home consumption and garden cultivation subsequently came together through group work parties and cross-visits to develop a more ambitious, watershed level agenda identifying water sources, vulnerable areas and conflict zones. A management plan was developed, that was turned into a project proposal that was later funded by the municipality.

Despite the evident success of the project in helping farmers to start addressing climate change, the project acknowledges some obstacles:

> Development practitioners need to be strong facilitators with flexible programmes and funding to support open-ended learning-action, which goes against the grain of standard (pre-planned projects and technology transfer). Donor and development agencies must hand over more trust and responsibility to communities to design and implement their own agendas. Local people and outsiders need to be free to learn from each other and to learn as they go along.
>
> (IIED, 2009: 60)

Source: compiled from IIED (2009).

method-oriented mania' has emerged that could lead to the same standardised approaches and solutions that they aimed to challenge. There are similar concerns that participation and empowerment have become such widely pronounced policy goals (and are often a condition for funding of research and development projects) that many are now engaging with participatory methods 'despite a superficial understanding of the underlying empowerment principles that were at the root of much pioneering work' (ibid.: 5). Clearly, ongoing critical reflection on complex questions of power in participatory research and development are required (Cooke and Kothari, 2001).

Secure rights and gains

Ensuring that individual land users and communities have secure rights to resources and the benefits from investments is a further characteristic of more sustainable rural development processes, identified in Figure 4.15. In short, farmers who sense their rights are insecure are deterred from taking a long-term view and from investing their labour and other assets further. When people are sure, for example, that they have the rights to the products from trees that they plant, invest in and manage, they plant many more than they do when there are restrictions on the use or harvesting of such resources (Chambers, 2008). Because of the centrality of agricultural production in rural livelihoods, issues of land ownership and tenure are central to the question of security of rights to resources and benefits. As well as being the foundation for economic opportunity, land is also regularly a major correlate of social prestige and political power in rural societies of the developing world. Yet substantial inequality in the distribution of landholdings is a widespread, entrenched and often worsening feature of many developing countries despite decades of land reform programmes (see Potter et al., 2008; Jacobs, 2010). Furthermore, many people (particularly in Asia) are engaged in agricultural production under systems of tenure including share-cropping arrangements, which deliver the major benefits to the landlord rather than the tenant. The limited autonomy of farmers under many forms of contract farming is also a concern, when not only does the contractor control many of the on-farm operations, but farmers can ultimately be evicted from lands if leases are contravened (Watts, 1994).

Through the late twentieth century, pressure for land reform in countries including South Africa, India, Bolivia and Brazil has increasingly come from grassroots movements. As considered in previous chapters, what may start as locally based protest around specific issues are increasingly finding commonality and support as part of transnational environmental justice movements and are influencing many areas of national policy and the decisions of international organisations. The shared concerns of these movements include exposing the social and environmental outcomes that occur when people lose their rights to land and to jobs and are mobilising to find more just and thereby sustainable processes. Brazil has one of the most unequal distributions of landholding in the world. In a country with a population of approximately 190 million people, just 35,000 families control half of all farmland and 3 per cent of the population owns two-thirds of all arable land (Zobel, 2009). Box 4.8 considers the work of one of the largest social movements in Latin America, the Movimento do Trabalhadores Sem Terra (MST), and its role in pressuring for new land reform legislation in Brazil. It also confirms how the movement has supported wider access to land through occupations and the formation of co-operatives and through providing such things as education and legal advice.

Box 4.8

Pressures for land reform in Brazil

The Brazilian landless workers movement, the Movimento do Trabalhadores Sem Terra (MST) was formed in 1984. This was one year before the end of the country's military dictatorship and the return of civilian rule. It began as a coalition of individual squatter groups primarily located in southern Brazil. Its guiding slogan remains 'land for those who work it' and the movement centres on challenging existing land tenure regimes that allow land to be taken from the poor, for and by the wealthy. This includes taking direct action to support squatter groups and substantial lobbying of political parties to introduce legislation and meet the commitments of the constitution. MST activities have widened to include the concerns of both rural and urban poor. It is considered to be one of the largest and most well-organised grassroots social movements globally.

MST has over 2 million members, has occupied thousands of hectares of idle farmland and won land titles on behalf of more than 350,000 families. Families come together in brigades and coordinators are elected from each state to form a governing body. When land and title have been secured, members have also established cooperative farms, organised systems for health care and education, built schools, clinics and even a university (the Florestan Fernandes National University in the state of Sao Paulo). A dairy cooperative formed by 550 MST families in the southern state of Santa Catarina is now the biggest milk producer in the region. However, the movement has suffered substantial opposition within the Brazilian media, from wealthy landowners and their supporters in Government and business. As many as 1,500 people have been killed in land conflicts. Two separate massacres of landless squatters in 1996 brought international and national attention to the plight of the rural poor in Brazil.

In 1998, President Cardoso announced a new programme of agrarian reform aimed at reducing the social tensions around land and enhancing access to land by the poor. This is widely recognised to have been as a result of rising populist pressure, in particular from the MST. It was a 'market-led' policy (with World Bank finance) that was designed to reduce state involvement in the implementation of land reform and to stimulate the market in land. In other words, the thinking was that separating land reform from state involvement and allowing price to determine how land is allocated would serve to improve the transparency of the process and in turn decrease support for the MST. Loans were made available to people who had previously worked in agriculture and had incomes of below US$15,000 to help them purchase land. People had to form associations with other interested buyers to purchase land from existing farmers. However, from the outset the programme generated significant dissent from MST. Targets for land transfers were low and were not achieved. Land prices also rose steeply through the period of the programme as activities of agri-business expanded. Not only did this give problems of land supply for the programme, but the costs of re-payments on loans also rose, creating problems of indebtedness. An estimated one-third of families' loans were in default in 2007. Progress on land reform faded further through the decade of President Luiz da Silva. This was despite optimism for his term given his working-class background and previous founding involvement in the Workers' Party. In 2005, the MST organised the biggest march in Brazilian history to push for implementation of the promises of the constitution that land left unproductive for two years is subject to agrarian reform.

Source: compiled from Zobel (2009); Wolford (2008); Sauer (2009).

For Middleton et al. (1993: 124), land reform is one of two 'inseparable starting points in any fight for sustainable agriculture', the other being the rights of women. As identified above, there are problems of data due to women's work in agriculture often being unpaid, undervalued and often not recorded; however it is estimated that 80 per cent of food in sub-Saharan Africa is produced by women (ibid.) and there may be twice as many women as men currently working in agriculture-related activity in developing countries (Momsen, 2010). Yet women own less than 2 per cent of the world's total agricultural area and women have often been excluded from programmes of land reform (Jacobs, 2010). Clearly, these patterns illustrate that questions of security of rights and gains for sustainable rural development include not only 'ownership' of resources (as encompassed in formal land laws, for example), but rights to control the products of labour that are defined by gender relations. In many parts of the developing world, women receive rights to land through their husbands on marriage, but may have few rights to decide what is cultivated or marketed or how any profits made are spent. Furthermore, there is often a gender bias in terms of access to training (such as technical assistance from extension workers) and access to inputs (including water and credit), that ensures that many agricultural development interventions have very different impacts on women and men (Momsen, 2010).

The role of women in conservation

It has been long and widely argued that women are key to sustainable development: 'the achievement of sustainable development is inextricably bound up with the establishment of women's equality' (WRI, 1994: 43). Early research and development work had focused on women's role in the domestic sphere and on bringing women into development through programmes addressing 'women's areas' such as family planning and nutrition (understood as the 'Women in Development' school of thought, WID). Through this, however, a better understanding of women's additional roles in agricultural production and wider productive activities, such as fuelwood and water collection and management, developed. By the 1980s, understanding rose of women's 'substantial interest' in environmental resources and of the many ways in which women were taking action on environmental management challenges. Ecofeminists took the notion of women's interest in environment further to suggest that women have a natural affinity with nature aligned to their child-bearing qualities that men do not have. Authors such as Vandana Shiva (1989), for example, have contrasted what they consider women's essential features of caring and empathy with the controlling, manipulative position of men, to argue that women may be the key to new and more sustainable ways of living and social relations, if these 'feminine principles' could be sufficiently recovered.

Through the 1990s, many 'Women Environment and Development' projects sought to work with women in this role as 'privileged environmental managers' (Braidotti et al., 1994) and to 'harness' their knowledge of local environments and their existing initiatives within diverse conservation projects. Whilst there are many

Plate 4.4 *Women in environmental management*

a) Fuelwood collection, Zimbabwe
Source: author.

b) Organising the community:
a Lampungese wedding
Source: Becky Elmhirst, University
of Brighton.

Plate 4.4 *. . . continued*

c) Water management, Malawi

Source: Huw Taylor, University of Brighton.

successful examples across soil and water conservation, in forestry and irrigation projects, there have also been problems (see Momsen, 2010 in this series). For example, the suggested synergy between women and the environment whereby they are the main 'users' and 'managers' of natural resources is now seen by many as reflecting accepted gender roles and a lack of alternative opportunities for women rather than any chosen or inherent empathy with conservation and natural resources. In other words it is through the gendered division of labour (whereby women take primary responsibility for collecting and managing the energy and water needs of the household) that women's immediate 'practical gender' needs (Pearson, 2001) regularly centre on accessing environmental resources. This in turn explains how environmental degradation will often impact most heavily on women, as seen in Figure 2.18. Accepted gender roles also structure how women tend to take greater responsibility for maintaining communal services such as schools, health posts (and community woodlots and wells) and generally often provide the 'glue' between the elements and activities of the community (such as organising weddings). The outcome of many

WED project interventions was that conservation became the 'fourth burden' on women to add to their gendered responsibilities in production, in the reproduction of the household and in community management, particularly through the way that they assumed that women had the spare time to undertake these projects.

Work within the 'gender and development' (GAD) school of thought now considers that it is not sufficient to work only with women or to assume that women's relationships with the environment are undifferentiated. Gender divisions of labour are not uniform or static, nor are gender relations. Furthermore, women are not an undifferentiated or homogeneous group. Relations between men and women may vary in such things as class or other social divisions, and are continually 'struggled over' (see Momsen, 2010). Box 4.9 illustrates the importance of working for women's needs in resource management and the kinds of resistance met, but also the importance of changing relations between women and men and between local villagers (men and women) and wider political structures for the sustained development of key resources in the livelihoods of the poor.

Box 4.9

Building women's rights in sustainable water management

The Self-Employed Women's Association (SEWA) was formed in 1972 as a small membership organisation for poor women working in the informal sector. It is now a trade union with over 1.2 million members across India. It is both an organisation and a movement in that it aims to empower poor, illiterate and vulnerable women through ensuring full employment and working towards income, food and social security. An estimated 200,000 of its members are poor, self-employed women working in the informal sector in the rural districts of Gujarat state.

In Gujarat rainfall is low and erratic and the majority of villages have no reliable source of fresh water – many relying on mobile water tankers that may or may not arrive. Women typically spend up to six hours a day collecting water. Agricultural productivity and pastoral production are compromised by drought and increasing salinity and there is substantial 'distress migration' of people during the summer months.

As a trade union, SEWA's work is mandated by the members themselves. It has a national executive formed of representatives of 'local associations' from the districts. At the village level women are organised into unions and cooperatives, often according to the different occupations that they are involved in.

In 1986, the State Water Board in Gujarat was aware of the failings of centralised management of local water resources in the districts. It approached SEWA, knowing their expertise in grassroots development, to look into involving local communities in building, operating and managing a water supply system. On the basis of its previous experiences, SEWA understood that it would be easier to recruit members into water development activities if these could be linked to economic improvement. SEWA mobilised women around eight economic activities ranging from embroidery to anti-desertification measures. By 2000, nearly 200 groups have been formed under the 'Development of Women and Children in Rural Areas Programme' – a joint initiative of the Indian government and UNICEF.

Historically, water infrastructure such as bore holes and dams was regarded as male territory. Many women were therefore often reluctant to come forward into this area and men were widely critical of women entering this public domain. Despite this, many of the water user committees established in Gujarat (*pani samities*) were all women or at least equal in number to male members.

SEWA has initiated many different types of activity amongst the water committees in various districts. These include visits to other functioning cooperatives to see how democratic frameworks operate and meetings with water engineers to understand water supply systems. Much of their work has involved assisting women to understand and negotiate their way through the maze that is India's system of governance of water resources. There are seven bodies solely at the national level, for example, that have some authority and responsibility over water. This gets more complex as it moves to district and sub-district and then to local *panchayat* levels.

In order to take steps to restore traditional sources of water, women had to learn about the roles of different agencies, decide which to approach and how. SEWA assisted in bringing engineers from the Minor Irrigation Departments together with the villages to plan, design, source and pay for materials.

The continual training, support for women to deal with the technical, institutional (and social and cultural demands) of water related activities has been fundamental to SEWA's success in securing the sustained participation of villagers. It has also been of paramount importance for institutionalising grassroots governance in the water sector whereby new institutions dominated by women have been created with strong links to existing governing institutions at wider levels. A shift in attitudes towards women has also been identified: they have earned respect within their families and their communities for their knowledge and abilities and mainstream institutions are now willing to accept illiterate women on their training programmes, for example. There have also been cases of districts abandoning contracts with private sector companies in favour of local organisations.

Source: compiled from WRI (2003); www.sewa.org

Assuring and securing rights to resources for women and confronting fundamental gender inequalities are continued challenges for sustainable development. Women may not necessarily be the source of solutions to environmental problems: groups of women as well as men may act in environmentally damaging ways and women's apparent 'closeness' to nature may stem more from their lack of opportunities outside the household and the social relations in society than their biological makeup. Sustainable development, therefore, depends on removing women's subordination and oppression as well as their poverty, not simply 'grafting on women as a group' or 'lumping' women's varied interests together (Middleton et al., 1993).

The capacity of local institutions

When people have secure rights to resources and gains from investments, it has been found that they also have a perceived self-interest in project development and implementation, and this is a further characteristic of more successful sustainable rural developments, as identified in Figure 4.15. Experience suggests that when people participate for the reason that they have seen success achieved and have become enthusiastic about achieving it for themselves, projects tend to be more relevant, to spread more quickly and to encourage further innovation and development. This was seen in Box 4.7 where visits to particularly innovative farmers in surrounding areas were important in inspiring other individuals to undertake soil conservation and water harvesting measures. Furthermore, as Pretty and Ward (2001: 209) suggest, 'for as long as people have managed natural resources, they have engaged in forms of collective action'. Such collaboration is evidenced in a huge variety of local associations that exist in rural areas across developing regions, such as work teams, burial societies and savings groups as well as in the long-standing 'traditional' social norms, rules and values that shape resource management and guide the association between individuals. Together these comprise important 'institutions of governance' that are the focus of much interest towards fostering more sustainable patterns and processes of development. In short, the importance of strong local institutions in sustainable futures is now understood in terms of their potential to foster more sustainable development innovations but also in that it can be the breakdown of such institutions that is part of the explanation for environmental degradation.

It was seen in Chapter 3 that non-governmental organisations in general and as one component of civil society have become more prominent in development thinking and action, especially as desires for democracy and justice have mounted within certain countries. NGOs are considered to have a number of characteristics that make them particularly suited for working towards sustainable development, including their tradition of working at the local scale, with people in specific environmental contexts, and for how they tend to be less bureaucratic than other (particularly governmental) organisations. It is also considered that NGOs tend to attract staff who themselves have the kind of qualities and values suited to working for and in more sustainable ways. As Chambers (1988: 13) suggested, staff of NGOs have historically been recruited for their 'sensitivity, insight and competence' and where the 'reversals of normal values are often most at home'. However, it has also been seen that NGOs are now working in new ways and in new partnerships (including with organisations that they may have historically challenged) such that their capacity to continue to work according to these values and in these ways may be compromised. Box 4.10 illustrates three projects in India that are all seeking to improve farmers' access to information for more secure rural livelihoods, based on new networks and partnerships between community-based organisations, national NGOs, state authorities and national and global private sector organisations.

The rise of civil society is also based on the importance now being given to 'social capital' in explaining and fostering development. Whilst a difficult concept to define (see Figure 4.19), many international organisations and donors 'found community' particularly in relation to natural resource management in the late 1990s (Agrawal and Gibson, 1999). In essence the thinking is that the assets on which people draw for their livelihood include 'social' resources such as through memberships of groups, access to networks and through 'informal' relationships of trust and exchange, for example. Social capital enables mutually beneficial collective actions (such as those required to manage 'common property' resources such as grazing or water courses). Social capital can also be enhanced by networks that improve people's ability to work together or to access wider existing institutions and resources, hence its potential for new approaches to conservation and development. In 2001, Pretty and Ward estimated that over 400,000 new community-based groups and networks worldwide may have been formed since the 1990s towards building social capital in natural resource sectors and largely in the developing

Box 4.10

Public–private collaborations for enhancing information and communication for rural development in India

The organisations and programmes:

1. The Indian Farmers Fertilisers Cooperative (IFFCO) is a national organisation of rural co-operatives working in partnership with a private-sector Indian mobile phone company to deliver a farmers' information service known as IFFCO Kisan Sanchar Limited (IKSL). The service requires farmers to purchase a SIM card through which they receive free voice-mails containing agricultural information (including on weather, market prices, fertiliser availability, electricity timings and government schemes) and have access to a helpline (costing 1 Rupee per minute).

2. Reuters is an established multinational private company providing information services globally. 'Reuters Market Light' (RML) is an India-based business providing mobile-enabled information to farmers. Farmers purchase a subscription (on a 3-, 6- or 12-month basis) to receive daily agricultural information (as in (1) above) through text messages.

3. MS Swaminathan Research Foundation (MSSRF) is an NGO that is piloting mobile-information services for fishermen in partnership with Tata Teleservices (an Indian mobile phone operator) and Qualcomm (a global technology company). The programme 'Fisher friend' provides free mobile handsets to fishermen (which they must share on a rotating basis with others in the community) and access to the project information service that includes weather, market prices, the optimal fishing zone and details of government schemes.

Impacts of the mobile communication and information:

1. Saving crop loss and improving yields: Mr Jagdish, Khanvaas village, Rajasthan.

Service: IKSL

Saving crop loss: this farmer acted on timely weather information received through IKSL to protect a harvested crop (*Gwar* – used as livestock fodder) that was lying on ground exposed to the elements. He estimates that he would have lost 50 per cent of this crop if not for this advanced information, worth Rs 5–6,000.

Increased revenue: the farmer made use of information provided by IKSL concerning planting techniques and diseases to make changes in farming practice. In his description, he shifted from 'guess-based' actions to following modern scientific cultivation practices, resulting in his estimation of a 25 per cent increase in annual earnings of Rs 25,000.

2. Optimising supply to increase revenue: Arphal village, Maharashtra.

Service: RML

The farmers in this village had been involved in horticultural cultivation for two years. Flowers are a highly perishable commodity and farmers monitor production and harvesting closely to minimise waste. The farmers received information from RML about a predicted increase in market demand for their crop. They applied a particular fast growth tonic to increase production and capitalised on the information through higher sales. The farmers reported that the amount of daily supply taken to market typically was less than 1,000 sticks a day. They have now adjusted the quantity they bring to market as a result of RML market demand information offering the potential for further increased revenues on high demand days.

3. Getting a higher catch: Mr K. Prabhakaran, Veeramptattinam village.

Service: Fisher friend

This fisherman had stayed on land to manage family commitments and was advised by colleagues at sea that they were having a poor fishing day. He told them about the optimal fishing zone information he accessed on his mobile phone and they quickly changed their location and benefited from a higher catch. One of the beneficiaries hauled Rs 30,000 on that day, between six and ten times the typical daily revenue reported by other fishermen with launch boats.

Source: compiled from Vodafone (2009).

Figure 4.19 *The essence of social capital*

The social assets comprising social capital include the norms, values and attitudes that predispose people to cooperate; relations of trust, reciprocity and obligations; and common rules and sanctions mutually agreed or handed down. These are connected and structured in networks and groups.

Source: Pretty (2008).

world. Examples of programmes established in specific countries are shown in Figure 4.20. These groups may be formed as new associations between different actors (such as between the state and community groups in joint forestry management), to build on existing community groups (for example, women's associations becoming more formalised in the context of micro-finance projects) or through identifying and strengthening existing informal institutions (such as enhancing traditional community organisations operating to manage common property resources like grazing lands). Clearly, the existence of new institutions does not ensure sustainable resource management although there has been much success in terms of farmers organised within groups performing better than those who are outside particular resource-based project initiatives (Pretty and Ward, 2001).

Figure 4.20 *Social capital formation in natural resource management, selected countries*

Watershed and catchment groups
- *India:* programmes of state governments and NGOs in Rajasthan, Gujarat, Karnataka, Tamil Nadu, Maharashtra, Andhra Pradesh
- *Brazil:* 275,000 farmers in three southern states adopted zero-tillage and conservation farming as part of watershed groups
- *Kenya:* Ministry of Agriculture catchment approach to soil and water conservation
- *Honduras/Guatemala:* NGO programmes for soil and water conservation and sustainable agriculture
- *Burkina Faso/Niger:* water harvesting programmes

Irrigation water users' groups
- *Sri Lanka:* Gal Oya and Mahaweli authority programmes
- *Nepal:* water users' groups as part of government programmes
- *India:* participatory irrigation management in Gujarat, Maharashtra, Tamil Nadu and Orissa
- *Philippines:* National Irrigation Administration turned over 12 million hectares to local management groups
- *Pakistan:* water users' associations in Punjab and Sindh Microfinance institutions
- *Bangladesh:* Grameen Bank nationwide
- *Pakistan:* Aga Khan Rural Support Programme in Northern Areas
- *Nepal, India, Sri Lanka, Vietnam, China, Philippines, Fiji, Tonga, Solomon Islands, Papua New Guinea, Indonesia and Malaysia* have a wide variety of bank and NGO programmes

Joint and participatory forest management
- *India:* joint forest management and forest protection committees in all states
- *Nepal:* forest users' groups

Integrated pest management
- *Indonesia:* 1 million graduates trained in rice and vegetable IPM programme
- *Vietnam, Bangladesh, Sri Lanka, China, Philippines, India:* a further 8,000 trained

Farmers' groups for research and experimentation
- *Kenya:* organic farming groups
- *Columbia:* farmer research committees

Source: compiled from Pretty and Ward (2001).

There has been close interest in the potential of so-called 'Community Based Natural Resource Management' (CBNRM) initiatives for more sustainable rural development. CBNRM encompasses a breadth of policies and programmes in practice, but typically involves one of three kinds of experience: some kind of joint or collaborative management involving local communities in the management of (typically) previously defined 'state' resources; those that look to decentralise authority to (usually newly created) local/community institutions; and those that work to strengthen the traditional, local institutions and controls that are already in place (Elliott, 2002). CBNRM initiatives are being undertaken for various reasons and can encompass quite divergent ideas about conservation:

> At one extreme fall existing conservation projects (e.g. conventional protected areas) that belatedly make minor efforts to draw in local people . . . at the other extreme lie initiatives aimed specifically at the development of particular (often 'sustainable') uses of natural resources by local people who are given full tenure over those resources . . . The first is based on the idea that conservation has to do with concern for 'wild' species and their associations (ecosystems and habitats) . . . the second is based on the idea of conservation as the sustainable management of renewable resources.
>
> (Adams and Hulme, 2001: 14)

Whilst high levels of donor support continue to go to CBNRM initiatives, there remain challenges, for example financial sustainability once donor funds end. Many traditional institutions in resource management also continue to be eroded through such things as processes of resource scarcity, social changes associated with migration and a decline in the authority of traditional leadership. Perhaps most fundamentally, many CBNRM initiatives (as in wider forms of participatory development considered above) have assumed distinct 'communities' exist, whereas they are in fact highly diverse and dynamic, contain many different interests and environmental priorities within them and have very varied abilities or power to negotiate and make change.

Conclusion

For some time, the debates about both the environment and development have been dominated by the interests and values of the rich rather than the poor, men rather than women, and the urban rather than the rural. The challenges of reversing these priorities have been seen within this chapter to be wide and substantial, yet integral to the prospects of sustainable rural development. Whilst individual farmers ultimately make many of the land-use and resource-management decisions in rural areas, ensuring that these are sustainable depends evidently on many broader factors. Enhanced integration into the global agri-food system has brought new risks for farmers as they engage in new kinds of relationship with substantially different kinds of organisation and wider economic markets, for example. The capacity of national governments to shape national agricultural policy and to support local farmers is also more constrained. The chapter has shown the benefits for environment and development that have been achieved through research and development initiatives that start with the local context, the priorities and knowledge of small-scale rural farmers, strengthen their rights over resources and networks of association, for example, and that evolve with learning and through participatory processes. But there are substantial remaining challenges including that 'community participation' is itself 'conflictual' (Mohan, 2002: 51) in that it challenges existing structures of power ranging from the legal authority of state institutions to accepted norms of behaviour such as gender roles and relations within the household. There is also the danger that focusing on grassroots civil society can divert attention away from the many processes affecting individual lives that cannot be tackled at the local level. However, the long-term benefits of providing secure rural livelihoods for the rural poor will also fall to wider communities including in urban centres.

Summary

Despite rising urbanisation, rural areas worldwide are expected to provide a range of functions central to sustainable development.

Rural development has been substantially 'left out' of development thinking and practice in the past. Recognition of the complex global challenges of sustainable development in the twenty-first century has led to a resurgence of interest in agriculture in particular.

Agriculture has become increasingly globalised. A key sustainability concern is the degree of control that major agri-business companies now have through many stages of the global agri-food system.

The components of rural livelihoods are diverse and change over time. Non-farm economic activities in rural areas are growing but agriculture remains important for the majority of rural households.

A 'Gene Revolution' for world agriculture is offered as a route to further increases in global food production, but there are many sustainability concerns and much opposition to the further use of GMOs.

Putting Farmers First has become a powerful narrative in rural development and has done much to guide more sustainable rural development interventions in practice. The chapter details what this means for research, for planning processes and for projects on the ground and the continued challenges revealed through these experiences.

Women often have a close relationship with environmental resources, but their 'interest' in environmental conservation can be theorised in different ways. Successful sustainable development initiatives centre on building women's rights in resource issues but also on changing gender relations within the household and with wider authorities.

There are many initiatives in natural resource sectors that look to strengthen the networks and associations of people within communities and with wider institutions including NGOs, the state and the private sector. Lessons are emerging regarding the challenges of community participation in practice and the contexts in which local institutions can thrive or break down.

Discussion questions

- Identify the principal differences between the 'Green' and 'Gene' Revolutions in world agriculture.

- Do women hold the key to sustainable rural development?

- How and why is agriculture only one aspect of rural livelihoods?

- What challenges for sustainable development are presented by the dominance of multinational companies within the global agri-food system?

Further reading

Conroy, C. and Litvinoff, C. (1988) *The Greening of Aid: Sustainable Livelihoods in Practice*, Earthscan, London. A key publication that drew together the early principles emerging of more sustainable development interventions in practice.

Lynch, K. (2005) *Rural–Urban Interactions in the Developing World*, Routledge, Abingdon. A clear introduction to how development processes and patterns in rural and urban contexts are inter-linked.

Rigg, J. (2007) *An Everyday Geography of the Global South*, Routledge, Abingdon. An excellent book that prioritises the everyday experiences and strategies of individuals and households and draws on a very wide number of case studies within countries of the developing world.

Scoones, I. and Thompson, J. (eds) (2009) *Farmer First Revisited: Innovation for Agricultural Research and Development*, Practical Action Publishing, Rugby. A critical reflection on the achievements and limitations of the original 'Farmer First' approaches to rural development.

Unwin, T. (ed.) (2009) *ICT4D: Information and Communication Technology for Development*, Cambridge University Press, Cambridge. A good source for understanding the ways in which new information and communication technologies are bringing challenges and opportunities in development.

Websites

http://www.steps-centre.org   A research organisation (Social Technological, and Environmental Pathways to Sustainability) located at the Institute of Development Studies at the University of Sussex that has a long-standing reputation for leading new ideas in rural development thinking and practice.

www.mstbrazil.org   English language website of a network of individuals and organisations working to support the Brazilian Landless Workers movement (MST).

www.iied.org   Website of the International Institute for Environment and Development, which has done a lot of work in the past on participatory learning and action and produces a host of news, briefings and research reports related to natural resources and sustainable development.

www.sewa.org   For details of the activities of the Self Employed Women's Association – a trade union that works to support poorer self-employed women throughout India.

5 Sustainable urban livelihoods

Learning outcomes

At the end of this chapter you should be able to:

- Realise that levels of urbanisation are rising but there are key differences between regions and within countries
- Be aware of the strong links between urbanisation and socio-economic development, but also that rising economic prosperity is a key factor shaping the environmental burden of cities
- Understand the factors underpinning the rising informality of urbanisation in the developing world
- Identify the key features of the urban environmental challenge at a number of spatial scales
- Understand how and why taking decisions as close as possible to urban citizens is understood as key to more sustainable urban developments

Key concepts

Urbanisation; green and brown urban environmental agendas; urban informality; slums; decentralisation; good governance

Introduction

In 1800 only 3 per cent of the total world population lived in towns and cities. In 2010, this figure was 50.6 per cent (UNHSP, 2010). Figure 5.1 confirms the trend worldwide for the increased concentration of people in cities rather than rural areas. Whilst such urbanisation occurred in Europe, North America and Latin America through the mid twentieth century, the trend is now increasingly being seen in developing regions and particularly within Asia and Africa. Whilst Latin America is the most urbanised region in the developing world, Asia has the largest number of people living in cities (over 1.5 billion people) and it is in Africa that some of the

Figure 5.1 *Proportion of urban population by region, 1950–2030*

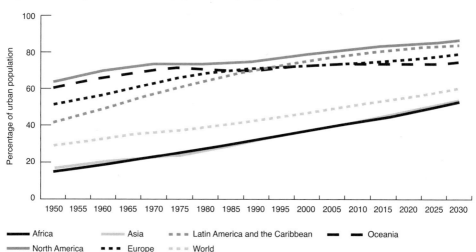

Source: UNHSP (2006).

fastest annual rates of urban growth are found (exceeding 4.5 per cent in some countries). Currently, there are more people living in cities of the developing world than in cities in more developed regions and it is predicted that 80 per cent of the world's urban dwellers will live in cities of the developing world by 2030 (UNHSP, 2010).

In 1987, the World Commission on Environment and Development suggested that the urban challenge lay 'firmly in the developing countries' (WCED, 1987: 237), due in the main to the unprecedented growth rates being observed and the challenge of meeting the immediate needs of an expanding urban poor. In that year the World Bank had estimated that approximately one-quarter of the developing world's absolute poor were living in urban areas (World Bank, 1990). By 2000, it was increasingly recognised that whilst urbanisation (and globalisation) are powerful forces of economic growth, innovation and creativity, they are also drivers of exclusion and environmental degradation. At the time of the Millennium Declaration for example, an estimated one-third of all city dwellers were living in slums. Improving the lives of slum dwellers was recognised as a major development challenge (rather than an unfortunate consequence of urbanisation). Addressing global poverty is at the core of the Millennium Development Goals, as seen in previous chapters, and achieving the goals is understood to be heavily dependent on how well cities perform:

The link between urbanisation and socio-economic development
cannot be disputed. Cities make countries rich. Countries that are
highly urbanised have higher incomes, more stable economies,
stronger institutions and are better able to withstand the volatility of
the global economy than those with less urbanised populations. The
experiences of developed and developing countries also indicate that
urbanisation levels are closely related to levels of income and
performance on human development indicators.

(UNHSP, 2006: 48)

There are substantial challenges for *all* cities in managing the
environmental implications of economic growth, in meeting the
needs of their residents and in protecting the environmental resources
on which they depend in the future in ways that are beneficial both
socially and economically for residents. City-based production
currently accounts for the majority of resource consumption and
waste generation worldwide, for example. As identified in Chapter 2,
cities consume more than two-thirds of global energy and produce
over 70 per cent of carbon emissions (World Bank, 2010a). Intra-city
inequality is also rising globally (UNHSP, 2010). The globalising
world economy also ensures that the nature and direction of urban
change in any city or country is also now more inter-linked and
integrated. Furthermore, with rising affluence, the environmental
burden of cities tends to fall increasingly at the global scale (and on
the future generations) confirmed, for example, in the higher per
capita emissions of carbon dioxide and levels of waste generation
from cities of the more developed world (McGranahan and
Satterthwaite, 2002).

The focus of this chapter, however, is the cities in the developing
world and the particular sustainability challenges therein that to a
large extent are not key concerns in wealthier cities. Box 5.1
describes and contrasts what have been termed the 'Green' and
'Brown' agendas for the most affluent and more impoverished cities
respectively. The level of affluence is not the only factor affecting
the environmental burden of cities, as will become clear through this
chapter. The differences identified in the box should also not detract
from the common, global challenges of sustainable urban
development. So too, the multiple deprivations of urban poverty
illustrated in this chapter should not deflect from the fundamental
pattern that affluence creates more pollution than poverty. In
addition, whilst the Brown Agenda is the priority for low-income
countries but a substantially 'old' issue for the more developed

Box 5.1

Green and Brown environmental agendas

The term 'Green environmental agenda', as shown in Figure 5.2, encompasses issues such as the depletion of water and forest resources. These concerns are most relevant to cities of the more developed world, to future generations and natural ecological systems. In contrast, the Brown agenda encompasses issues of access to basic water supplies, sanitation and housing; the 'pollution' of urban poverty that is most relevant to poor urban residents of today and to human health.

Figure 5.2 *The Green and Brown urban environmental agendas*

Features of problems on the agenda	The Brown Agenda	The Green Agenda
Principal impact	Human health	Ecosystem health
Timing	Immediate	Delayed
Scale	Local	Regional and global
Worst affected	Current generation	Future generations
Aspects emphasised in relation to		
Water	Inadequate access and poor quality	Overuse; need to protect water sources
Air	High human exposure to hazardous pollutants	Acid precipitation and greenhouse gas emissions
Solid waste	Inadequate provision for collection and removal	Excessive generation
Land	Inadequate access for low-income groups for housing	Loss of natural habitats and agricultural land to urban development
Human wastes	Inadequate provision for safely removing faecal material (and water) from living environment	Loss of nutrients in sewage and damage to water bodies from its release of sewage into waterways

Source: adapted/compiled from McGranahan and Satterthwaite (2002).

Whilst affluent cities can be considered to have performed better in terms of meeting the needs of their current populations, historically these have been met by displacing the environmental burdens over space (elsewhere) and time (delayed). For example, sewers have been put in to take human waste out of the city and goods whose production may have been resource intensive or damaging have been imported. The burden of dealing with the levels of waste generated or high levels of fossil fuel combustion will fall on the next generations through their contribution to global warming, for example. In wealthier cities, the key challenges for action lie in reducing excessive consumption of natural resources and the burden of wastes on the global environment (WRI, 1996).

In contrast, the environmental burdens of low-income cities are generally falling now and within the city. In addition, it is the most impoverished groups who suffer ill-health, vulnerable and shortened lives. The spatial shift in environmental burdens with urban economic development is modelled in Figure 5.3. As identified by McGranahan et al. (2001: 10):

> As we move into the twenty-first century . . . the importance of the 'brown' agenda is undiminished . . . economic change is outpacing urban environmental management and the achievement of social justice. Moreover, there is a serious danger that as new 'green' concerns are added to the environmental agenda, the 'brown' concerns will be neglected or misrepresented.

Figure 5.3 *A stylised environmental transition*

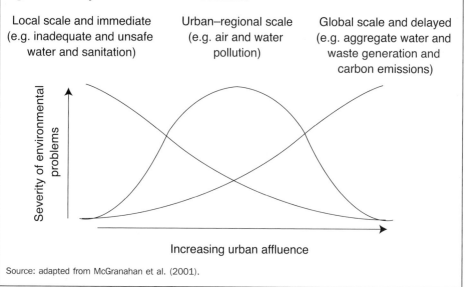

Source: adapted from McGranahan et al. (2001).

countries, the local and immediate problems in the developing world are bound up with 'new economic, social and ecological processes many of which are global in scale' (McGranahan et al., 2001: 6) and will require actions beyond that scale, as seen in subsequent sections.

Evidently cities are central to attempts at meeting the goals of sustainable development in the sense that this is where the majority of the world's population is located, with all the associated physical demands (such as for food and shelter) and the impacts involved in meeting the political, social and cultural requirements of urban living and the adoption of urban values. However, cities also provide opportunities for more sustainable development. Figure 5.4 summarises the key environmental advantages that cities can provide at a number of scales. The large numbers and concentration of people and activities are evident opportunities for economies of scale (lower unit costs) in providing services such as piped water, roads, electricity, and can reduce the costs of providing emergency services, for example. The risks from natural disasters can also be reduced more cheaply and effectively in urban centres through measures such as drainage to reduce the risk of flooding and improved buildings to better withstand flooding when it does occur. Opportunities in urban centres also include the potential to reduce fossil fuel consumption through the increased provision and use of public transport rather than private motor vehicles and the enhanced scope for recycling and reuse presented where large numbers of people live in close proximity. The concentration of people in cities can also facilitate their involvement in local district- and city-level politics and partnerships; and in the same way, it is easier for authorities to collect charges and taxes for public services and to fund environmental management. But close investigation is needed to consider where these opportunities fall – to governments, private enterprises or particular groups, for example – and who is excluded or unable to access such benefits.

Figure 5.4 *The environmental advantages of urbanisation*

For **urban and peri-urban** environmental living conditions:

● Lower costs per capita of providing treated water, sewer systems, waste collection, clean fuels and many other environmental services.

● More possibilities for local governments to fund or manage other forms of infrastructure and services that reduce environmental health risks (e.g. enforce pollution control and occupational health and safety).

For **regional** environmental burdens:

● High population densities, reducing per capita demand for occupied land.

● A concentration of major polluters, facilitating pollution control.

For **global** environmental burdens:

● Compact urban settlement patterns, reducing transport distances, increasing opportunities for more energy-efficient public transport, and thereby reducing carbon emissions.

● Economies of scale and agglomeration, making electrical co-generation possible and facilitating the use of waste process heat from industry or power plants for local (neighbourhood) water and space heating, again reducing carbon emissions.

Source: McGranahan et al. (2004).

This chapter investigates the key features of these processes and patterns of urban development in the developing world to illustrate more fully the nature of the urban environmental agenda. As discussed in the previous chapter, there are many processes of change currently that make rigid distinctions of urban and rural problematic. Just as large numbers of rural people now engage in non-agricultural activities and indeed may commute seasonally or daily to urban areas, large sections of the population of 'urban' areas also work in agricultural enterprises or in industries that serve rural demand. Cities also have significant environmental linkages with rural areas – through the resources they draw on (such as in agriculture and in energy supply) and the pollution they generate (influencing air and water qualities in surrounding areas, for example). As seen in Chapter 4, these sectoral and spatial linkages as well as the substantial flows of goods, income, capital, information and people between rural and urban areas are central to understanding the challenges and opportunities of sustainable development. In short, increased urbanisation and urban expansion bring many changes in land use, household composition, and access to infrastructure, services and employment and pressures on natural resources both within urban and rural contexts and also in peri-urban areas. However, in continuity with the arguments made in the previous chapter, there are differences between these contexts that continue to be important in understanding the demands and prospects of sustainable development.

Key patterns and processes of urban change in the developing world

Whilst the general trend, as seen in Figure 5.1, is for increasing levels of urbanisation across the developing world, the figure also confirms significant differences between regions and countries in the patterns of change. For example, the populations of Africa and Asia remain predominantly rural. But projections suggest that these regions will reach the 'tipping point' whereby over 50 per cent of their populations are living in towns and cities by 2030 and 2023 respectively. Whilst Africa as a continent currently has relatively small numbers of people residing in urban centres, it experienced the highest levels of urban growth in the world between 2005 and 2010 with average annual growth rates in excess of 3.3 per cent (UNHSP, 2010). However, it is in Asia that the largest numbers of people currently reside in urban areas and where the greatest future expansion in terms of additional urban residents is projected to occur.

There is also substantial diversity within regions and within countries. Whilst high rates of urban growth continue to characterise urban change in developing regions generally, 'not all cities contribute equally to this rapid growth, and neither is it unprecedented or out of control' (UNHSP, 2010: 15). Current levels of urbanisation are twice as high in North Africa as they are in East Africa, for example (52 per cent and 24 per cent respectively). Cities such as Mumbai in India and Shanghai in China are currently (and have been for several decades) amongst the largest cities in the world. Whilst their total populations are large (20.1 million in the case of Mumbai and 15.8 million in Shanghai in 2010), their annual rates of growth are relatively modest (1.96 per cent and 1.7 per cent respectively). As a contrast, several cities in China such as Shenzhen and Shangqiu grew annually in the 1990s by over 17 per cent from much smaller population bases (UNHSP, 2010). Whilst the projections that most of the world's demographic growth in the next 30 years will be concentrated in urban areas (rather than spread more evenly in urban and rural areas as in the past) is unprecedented, on a global scale, the pace of urbanisation is declining. Urbanisation expanded at an average annual rate of over 3 per cent through the 1950s, but has slowed to an expected 1.9 per cent between 2010 and 2015 (UNHSP, 2010).

Some caution is needed when considering all such rates and projections of urban change. Fundamentally, what constitutes an 'urban' area is defined differently in each country, which affects comparability across regions and countries. In Uganda, for example, a settlement of over 2,000 people is classified as urban, whilst in Nigeria it is 20,000 (UNHSP, 2010). In China, those settlements with more than 3,000 residents are considered 'urban' but only those with more than 60,000 people are identified as 'cities'. As seen above, the rates of increase identified for Africa are generally expansions over a small base and growth is occurring more widely across many smaller and intermediate urban centres than is the case in Latin America, for example. Some of the fastest increases in Africa were after independence that heralded the removal of colonial restrictions on the rights of local African people to live and work in cities. Projections are also based on the historic association between economic growth and urbanisation, yet many countries (particularly in Africa) are not experiencing steady economic growth.

Furthermore, some cities are expanding through contemporary influxes of refugees that may constitute temporary growth.

Certainly, the growth rates predicted for some of the largest cities in the developing world in the 1970s did not materialise in subsequent decades. For example, in 1980 it was predicted that Mexico City would have 31 million people by 2000, yet in fact it only reached 18 million (Satterthwaite, 2008a). Clearly, the processes underpinning urban change will be specific to particular places and times. For example, the 'early' transition to urban living that occurred in Latin America and the Caribbean generally (relative to other developing regions) as seen in Figure 5.1 has been explained in terms of the very unequal agrarian structure (leading to little capacity to retain a rural population) and government policies favouring import substitution that created significant urban based industrial employment (UNHSP, 2010). However, in the case of the mismatch between predicted and actual growth for Mexico City, economic growth slowed through the 1980s and 1990s, so fewer people moved to larger cities, but decentralisation policies (considered in more detail and more widely below) also made more resources available to smaller cities and therefore encouraged the growth of smaller centres.

In short, new kinds of urban system are developing worldwide, which include networks of very dynamic, although smaller cities as well as urban configurations that have not been seen before (Figure 5.5). Some of the key processes of urban change in the developing world are also without historical precedent. For example, in nineteenth-century Europe, people migrated to the towns and cities in search of employment and economic advancement. The industrial activities located in those areas depended on this process of migration to raise output and generate wealth. Urbanisation, industrialisation and 'modernisation' (the adoption of urban values) were processes which occurred simultaneously in the cities of Europe and were mutually reinforcing. This has often not been the case in the developing world. Table 5.1 highlights the cases of a number of Latin American countries in the 1960s (a period of relatively rapid industrial development), where it is seen that employment growth lagged substantially behind that in manufacturing output. Such 'jobless growth' continued to be a feature of urban change in the developing world into the 1990s and is seen currently in the decoupling of urbanisation from economic growth and poverty reduction, most evidently on the African continent.

Figure 5.5 *Novel urban configurations*

Mega-regions: economic units that result from the growth, convergence and spatial spread of geographically linked metropolitan areas. They are polycentric urban clusters surrounded by low-density hinterlands and they grow considerably faster than the overall country population. Recent research suggests that the world's 40 largest mega-regions are home to fewer than 18 per cent of the world's population but account for 66 per cent of global economic activity and 85 per cent of technological and scientific innovation. Examples include:

> In China, the Hong Kong–Shenzen–Guangzhou mega-region that encompasses a population of approximately 120 million people
>
> In Brazil, the mega-region that stretches from Sao Paulo to Rio de Janeiro that is home to 43 million people

Urban corridors: where a number of city centres of various sizes are connected along transport routes in a linear development often linking a number of mega-cities and their hinterlands and often related to specific economic objectives. Examples include:

> In India, the industrial corridor developing between Mumbai and Delhi stretching over 1,500 km
>
> In Malaysia, the manufacturing and service industry corridor encompassing Kuala Lumpur, the Klang Valley conurbation and through to the port city of Klang
>
> In West Africa, the greater Ibadan–Lagos–Accra urban corridor spanning over 600 km across four countries

Urban corridors are changing the way that cities and towns function, creating new forms of interdependence amongst cities and leading regional economic growth. They can also lead to unbalanced regional development as they strengthen ties to existing economic centres rather than diffuse development.

City regions: City regions have grown enormously in recent decades as major cities extend beyond their formal boundaries to engulf and absorb intermediate city centres, smaller towns, semi-urban areas and their rural hinterlands. Some of these city regions are now bigger in surface area and population than whole countries. Examples include:

> The Bangkok region in Thailand that is expected to expand another 200 km from its current centre by 2020 and to grow far beyond its current population of 17 million
>
> In Brazil, Metropolitan Sao Paulo currently spreads over 8,000 square kilometres and has a population of 16.4 million
>
> South Africa's Cape Town city-region draws daily commuters from over 100 km

Source: compiled from UNHSP (2010).

Whilst Africa as a whole has some of the highest urban growth rates in the world, such urbanisation is suggested to have occurred *despite* poor economic growth, inequitable distribution of resources and rising urban poverty (UNHSP, 2010). These factors are exacerbated by high levels of unskilled labour and the HIV/AIDS pandemic, for example, all of which are both a cause of persistent poverty and further impede efforts to reduce poverty and enhance the sustainability of many cities in Africa. However, whilst many commentators are quick to suggest the 'exceptional' case of Africa,

Table 5.1 *Industrialisation and employment in selected Latin American countries, 1963–9*

Country	Manufacturing annual output growth (%)	Manufacturing employment growth (%)
Brazil	6.5	1.1
Colombia	5.9	2.8
Costa Rica	8.9	2.8
Dominican Republic	1.7	–3.3
Ecuador	11.4	6.0
Panama	12.9	7.4

many countries on the continent have had high levels of economic growth in recent years (fuelled particularly by the mining, commodities and construction sectors as well as in services). It should also be understood that even in Asia's most successful and rapidly industrialising countries such as China and India, there are persistent problems of mounting inequalities with national GDP rates rising more quickly than poverty rates are falling, for example.

The rising 'informality' of urbanisation

Securing employment is critical to individual urban livelihoods and for patterns of urban development that are more sustainable. In 1997, Todaro suggested that one of the most 'obvious failures' of the development process over the past few decades had been 'the failure of modern urban industries to generate a significant number of employment opportunities' (p. 247). Few of the urban poor can afford to be unemployed for any length of time. Many, in fact, will be under-employed; either they are working less than they would like or are doing so at such low rates of production that their labour could be withdrawn with very little impact on overall output. Opportunities for formal sector employment also declined under structural adjustment programmes as jobs in the public sector were cut under government austerity measures and as industries were denationalised, for example. Whilst developed and developing countries alike are currently experiencing an employment crisis as a result of the recent global economic downturn, the impact in developing countries may be particularly severe, where unemployment is endemic and in most countries, the working age population is growing much faster than gainful employment opportunities (UNHSP, 2010). All such factors have led to a growing 'informalisation' of the urban economy in many developing countries and particularly in Asia and Africa.

In response to a lack of employment opportunities within the 'formal' sector, many urban residents in the developing world look to a wide variety of both legitimate and illegitimate income opportunities available within the 'informal' economy. This term is commonly used to refer to small-scale, unregulated, semi-legal economic activities which often rely on local, internal resources, family labour and traditional technology. The host of activities encompassed within this sector is illustrated in Figure 5.6. Whilst

Figure 5.6 *Informal sector activities*

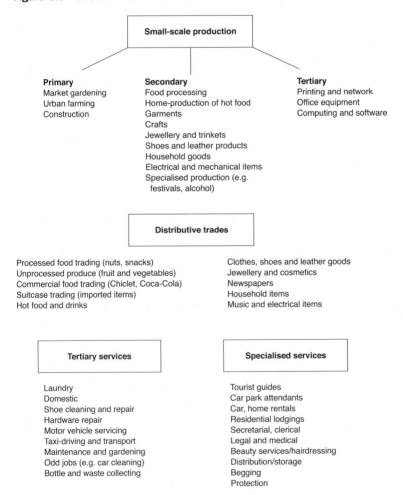

Source: compiled from Potter and Lloyd-Evans (1998); Drakakis-Smith (2000).

Figure 5.7 *Opportunities and challenges of informal sector employment*

Advantages	Limitations
More buoyant and elastic in generating jobs for an increasing urban labour force than the formal sector.	Low productivity of sector and its lack of bargaining power means incomes generally lower than in formal sector.
Small scale of operations and low levels of capital required lowers costs of creating employment.	Nature of employment means that earnings tend to be more intermittent and erratic, making access to formal credit mechanisms by households difficult.
Produces jobs that require fewer skills and less training than the formal sector.	Irregular and often illegal nature of many activities makes operators in the informal sector vulnerable to official and non-official harassment and persecution.
Lack of regulation and control and ease of entry makes informal sector well suited to absorption of migrants and other newcomers to the urban labour market.	Unregulated nature of informal sector makes it difficult for people to obtain access to services and supports necessary for increasing earnings and moving out of poverty.
Provides a safety net in times of economic crisis for those made redundant.	Informal nature makes it difficult to protect those who are engaged in them, whether as paid workers or as unpaid family members, against child labour abuses or against hazards in the workplace, for example.
	Informal sector jobs don't produce government revenues to support welfare policies, social safety net programmes.

Source: compiled from UNCHS (2001).

there are some problems with quality data, it is estimated that the sector provides employment, goods and services for as much as 60 per cent of the urban population. Recent research suggests that informal sector activities contribute between 80 and 90 per cent of all new job opportunities in Latin America and over 90 per cent in Africa (UNHSP, 2010). In New Delhi, approximately one-third of investment in housing has been from informal sector revenues generated by slum dwellers. Certainly, 'the extent and impact of poverty on urban populations, as well as on urban and national economies would be much greater were it not for the informal sector' (UNCHS, 2001: 212). The linkages between the informal and formal sectors of the economy and the contribution of the informal sector to national GDP are now better recognised (UNHSP, 2006). However, there are a number of challenges of employment within the informal sector, as shown in Figure 5.7.

Poverty is a common reality for many urban residents of the developing world. This is despite the generally positive relationship

Plate 5.1 *Urban informal income opportunites*

a) Door-to-door welding, Harare, Zimbabwe
Source: author.

b) Garment production, Kairouan, Tunisia
Source: author.

Plate 5.1 ... *continued*

c) Food trading/transport, Kolkata, India
Source: author.

between urbanisation and socio-economic development noted above, that is, that worldwide, countries with higher levels of urbanisation have higher levels of human development and the proportion of people in poverty is lower in urban centres than in rural areas. The absolute numbers of people living at or below poverty levels are currently higher in urban areas than in rural. By 2035, it is predicted that cities will become the predominant sites of poverty in terms of proportion as well as numbers. Furthermore, income inequalities are often more entrenched and apparent in cities than in the countryside and the cost of living in urban centres is also higher. Generally, urban dwellers have to purchase many items which can be accessed freely or more cheaply in rural areas, such as water, fuel and building materials. Estimates suggest that poorer urban households may spend between 5 and 10 per cent of their income on water (WRI, 1996) for example and may pay 30 per cent higher costs for food than their rural counterparts (UNCHS, 2001). The doubling of global food prices for many basic foods between 2007 and 2009 is understood to have impacted particularly strongly on the urban poor who spend upwards of 70 per cent of this income on food (UNHSP, 2010). Many goods and services are more commercialised in urban centres and urban residents are more reliant on cash income to secure these, which also brings a certain vulnerability to price rises and any drop in income. An understanding of urban poverty has become even more important in recent years as many basic services and housing, for example, are often no longer provided by the public sector but have to be accessed in the marketplace where people's ability to pay is critical. In recent years, there has been mounting evidence that the privatisation of public utilities including water and electricity (considered further below) has led to increased costs for consumers to levels where poor families cannot afford sufficient quantities to secure their most basic needs.

In continuity with the discussions of poverty in Chapters 2 and 4, the extent of urban poverty is unlikely to be captured by indicators based on income alone. Thinking only in terms of income can hide other aspects of deprivation such as poor-quality housing or people's capacity to challenge detrimental changes in their local environments. Figure 5.8 displays the multiple aspects of urban poverty and some of the wider factors underpinning these dimensions. Through this chapter, some detail of how these aspects of poverty impact on the prospects for sustainable development will be highlighted: the inadequacy of basic services and their links to good health and education, for example, and the significance of representative, democratic and accountable local authorities for ensuring people's rights to organise and for their protection against eviction will be seen. As Mitlin and Satterthwaite (2004: 12) state:

> of the multiple deprivations that most of the urban poor face, many of these deprivations have little or no direct link to income levels, while many relate much more to political systems and bureaucratic structures that are unwilling or unable to act effectively to address these deprivations.

Because of their poverty, many residents of cities in the developing world live in locations and settlements which are hazardous and detrimental to their own well-being. As noted in Chapter 2, as poverty becomes more concentrated into certain locations, often those of high ecological vulnerability, the urban poor may degrade these environments further in the course of securing their basic

Figure 5.8 *The deprivations associated with urban poverty*

Source: Mitlin and Satterthwaite (2004).

needs. Fundamentally, the poor are unable to afford the locations that are more desirable in terms of the inherent or acquired characteristics of the land. Wherever the urban poor are concentrated in cities of the developing world, it is commonly at high densities in areas of low rent. Poor groups do not live here in ignorance of the dangers; they choose such sites because they meet more immediate and pressing needs. These places are often the only ones where they can build their own houses or rent accommodation. The sites remain cheap because they are dangerous (Hardoy et al., 2001).

Regularly such locations are close to hazardous installations, such as chemical factories, and suffer continuous air and water pollution as well as the prospect of sudden fire or explosion. But critically for the urban poor, these locations may be close to jobs. As Gupta (1998) revealed, it was the high concentration of low-income people around the Union Carbide Factory in Bhopal which caused so many to be killed or permanently injured (over 3,000 dead and approximately 100,000 seriously injured). In Caracas, the capital city of Venezuela, the occurrence of slope failures has been increasing in recent decades. The maps of where these slope failures occur and those of the areas of low-income housing are almost entirely the same (Potter et al., 2008). Caracas grew rapidly in the second half of the twentieth century with the development of the oil industry and much of the expanded population were housed in self-built *barrios*, the majority of which are located on the steep slopes of the narrow east–west valley in which the city is located. Furthermore, whereas earthquakes were the trigger for past landslides, it is now heavy rains that regularly present a hazard to the vulnerable populations of the *barrios*. In 1999, many of those who died in catastrophic floods were low-income households living on unstable hill-slopes (Hardoy et al., 2001).

One of the most visible patterns of urban poverty in the developing world is the prevalence of 'slums', whereby large concentrations of low-income housing are seen on the fringes of many cities. Since 2000, there has been an international target to significantly improve the lives of at least 100 million slum dwellers within MDG7. This has become known as the 'Cities without Slums' target. However at the time of the launch of the MDGs, there was no operational definition of 'slums' or assessment of the number of slum dwellers globally. Many planners, city authorities (and slum dwellers themselves) had preferred less derogatory terms such as 'unplanned neighbourhood', 'squatter' or 'informal' settlement and different countries defined the physical and social attributes of these settlements differently. In 2003, the UN-HABITAT developed a provisional definition of a slum as a settlement in an urban area in which more than half of the inhabitants live in inadequate housing and lack basic services. To operationalise the definition, measurable indicators were needed at the household level. The definition of slums now used in UN monitoring is shown in Figure 5.9. The 2003 research found close to 1 billion slum dwellers in the developing world (approximately one-third of the total urban population) suggesting that the Millennium target was extremely

Figure 5.9 *UN-HABITAT slum indicators*

A slum household consists of one or a group of individuals living under the same roof in an urban area lacking one or more of the five amenities:

1. Durable housing (a permanent structure providing protection from extreme climatic conditions)
2. Sufficient living area (no more than three people sharing a room)
3. Access to improved water (water that is sufficient, affordable and can be obtained without extreme effort)
4. Access to improved sanitation facilities (a private toilet, or a public one shared with a reasonable number of people)
5. Secure tenure (*de facto* or *de jure* secure tenure status and protection against forced eviction)

NB: since information on secure tenure is not widely available, UN statistics use only the first four indicators to define slum households and then to estimate the proportion of urban population living in slums.

Source: UNHSP (2010).

modest. Indeed the MDG slum target has been achieved: 227 million people were moved out of slum conditions between 2000 and 2010, over half of whom were in China and India alone, confirming the low target that was set. Table 5.2 confirms that whilst substantial progress has been made across many regions (and particularly in Asia and North Africa), absolute numbers of people living in slum conditions continues to rise. Sixty-one million new slum dwellers were added to the global urban population in the decade from 2000, for example. Box 5.2 highlights the approach taken in Morocco towards cities without slums, or 'Villes sans bidonvilles' as it is known locally.

Whilst not all of the world's urban poor live in slums (nor are all slum dwellers income poor), the MDG target on slums has helped put the physical dimensions of urban poverty to the fore, that is, the deprivations of water, sanitation, shelter and overcrowding associated with urban poverty. The visible manifestation of housing that lacks basic services, space and security can take many forms. 'Stereotypical' forms of slums include the shanty towns often cramped together on the edges of cities but also dilapidated houses within city centres (typically one-story units that are visibly 'temporary' constructions and are often officially unauthorised). Less obvious forms of slum housing are the 'legal' dwellings that may look permanent from the outside, such as multi-storey public housing, tenements and dormitories, but where living conditions are seriously compromised through overcrowding, decay and poor upkeep. These are found on the urban periphery and within city centres. Figure 5.10 summarises a range of low-income options for

Table 5.2 *Urban population living in slums, 1990–2010*

Major region or area	*Urban slum population (thousands)*				
	1990	*1995*	*2000*	*2005*	*2010*
North Africa	19,731	18,417	14,729	10,708	11,836
Sub-Saharan Africa	102,588	123,210	144,683	169,515	199,540
Latin America and the Caribbean	105,740	111,246	115,192	110,105	110,763
Eastern Asia	159,754	177,063	192,265	195,463	189,621
Southern Asia	180,449	190,276	194,009	192,041	190,748
South-Eastern Asia	69,029	76,079	81,942	84,013	88,912
Western Asia	19,068	21,402	23,481	33,388	35,713
Oceania	379	421	462	505	556
Major region or area	*Proportion of urban population living in slums (%)*				
	1990	*1995*	*2000*	*2005*	*2010*
North Africa	34.4	28.3	20.3	13.4	13.3
Sub-Saharan Africa	70.0	67.6	65.0	63.0	61.7
Latin America and the Caribbean	33.7	31.5	29.2	25.5	23.5
Eastern Asia	43.7	40.6	37.4	33.0	28.2
Southern Asia	57.2	51.6	45.8	40.0	35.0
South-Eastern Asia	49.5	44.8	39.6	34.2	31.0
Western Asia	22.5	21.6	20.6	25.8	24.6
Oceania	24.1	24.1	24.1	24.1	24.1

Source: adapted from UNHSP (2010).

Box 5.2

Addressing the challenges of slum development in Morocco

In 2003, an assessment of housing conditions in urban Morocco established that approximately 5 million people and one-third of the total urban population were living in sub-standard housing, many of whom were urban slum dwellers. This was despite a decade of slum upgrading projects through the 1990s in recognition that informal housing developments were fast outpacing formal provision. In 2004, the government of Morocco initiated a comprehensive reform programme, to address issues of supply and demand for housing, to enhance the role of the private sector in housing provision, to improve the regulatory and institutional environment of the housing sector and to increase the affordability of low-income housing. A central part of the strategy was towards slum rehabilitation and clearance, known as the 'Villes sans bidonvilles' programme. Backed by a World Bank loan of US$150 million, the programme aimed to provide decent and affordable accommodation for 212,000 households across urban Morocco by 2010.

The programme was led by the Ministry of Housing and Urban Planning and included finance from World Bank, government and private sources. It was organised as a partnership between all stakeholders: developers, slum dwellers, municipal authorities and NGOs. The programme has involved:

● On-site upgrading of slum areas through the provision of roads, drainage and water supply, public lighting and electricity networks
● The provision of serviced plots on urban land on which households will build new dwellings
● The resettlement of slum households to newly built apartment buildings.

Whilst the first option is identified as the most beneficial and least disruptive for people, in practice the major part of the programme has involved the resettling of households away from poorer areas (typically areas of the more central Medinas) to newly built neighbourhoods on the edges of cities.

In the decade to 2010, the proportion of the urban population of Morocco living in slums declined from 24.2 per cent to 13 per cent (UNHSP, 2010). An estimated 2.4 million people have been moved out of slum conditions. Factors underpinning this success are considered to include the strong political leadership and pro-poor stance under the constitutional monarchy of King Mohammed VI; a centralised role of the Ministry of Housing that was also able to coordinate the large number of public authorities, private banks, NGOs and community groups involved; and sustained budgetary resources for the programme (UNHSP, 2010). However, the World Bank's Poverty and Social Impact Analysis of the programme of slum upgrading (World Bank, 2006) highlighted a number of issues including:

● A lack of knowledge within the programme of the preferences and diverse needs of slum-dwellers themselves that was a contributing factor in local resistance and limited participant engagement

● Insufficient participation of slum dwellers in the design and implementation of the housing programme

● Lack of affordability of apartments for the lowest income groups and the need for a more flexible approach to the housing design and supply

● A need for housing improvements to be integrated with programmes to enhance access to community and municipal services and employment opportunities if poverty alleviation goals are to be achieved.

Whilst many families appreciated their new-found legitimacy and ownership, the apartments do not provide sufficient space for extended families (important for the care of the elderly for example); they tended to be located at some distance from employment; and led to a decline in the potential for informal commercial activities. In some cases, economic and social hardship for those re-housed had increased (World Bank, 2006).

The government of Morocco has recently introduced legislation towards requiring greater public participation in all planning decisions and launched a national strategy of human development towards better integrated poverty solutions. In short, there is an acknowledgement that 'top-down' approaches to slum improvements can only go so far (UNHSP, 2006).

housing. The environmental impacts of living in such conditions are considered further in the sections below. Security of tenure and ownership is a key concern for both environmental improvements and for achieving poverty goals, yet it is currently difficult to monitor, as identified in Figure 5.9. In the 1960s and 1970s, for example, many city authorities had policies of slum or squatter clearance, where inhabitants had no defence in the law against eviction (Hardoy and Satterthwaite, 1989). Every year, it is estimated that several million people are evicted from their homes as a result of public works or government redevelopment programmes (UNCHS, 1996: xxviii). In cities including Mumbai, Jakarta and Rio de Janiero, despite the political commitments to (and international recognition of) pro-poor urban policies, thousands of people continue to be forcibly evicted from their homes 'all in the name of becoming a "world class city"' (ibid.: 165). Furthermore, because many settlements are illegal, they lack basic or emergency public services. Residents may also be ineligible for loans to improve their housing or employment conditions (their illegal shelter or land site being unacceptable as collateral), or indeed, for government subsidies such as in education (for which an authorised address may be required to register children).

In addition to the millions of people for whom accommodation is very insecure or temporary, there are also those who have no home at all and instead live on the streets. Again, official statistics are often not collected or can be misleading. Joshi et al. (2011) cite the National Census of Bangladesh in 1997 reporting only 32,081 homeless people nationally; a year later a study by the Asian Development Bank found 155,000 homeless people in Dhaka city alone. However, it is understood that both the homeless and the wider urban poor are a very diverse and heterogeneous group, which further confirms the complex challenges for sustainable urban development.

Figure 5.10 *The different kinds of rental and 'owner occupation' housing for low-income groups in cities of the developing world*

Type	Characteristics
Rented room in subdivided inner-city tenement building	Usually very overcrowded and in poor state of repair
Rented room in custom-built tenement building	Usually host many more families than built for. Often poorly maintained
Rented room, bed or bed hours in boarding house, cheap hotel or pension	Tend to be poorly maintained with a lack of facilities
Rented room or bed in illegal settlement	Share problems associated with illegality. Extra problems of insecurity of tenure
Rented land plot on which a shack is built	Highly insecure tenure; owner may require them to move at short notice
Rented room in house or flat in lower- or middle-income area of the city	Quality may be relatively good. May be located far from jobs
Employer-housing for low-paid workers	Quality often poor. Regularly rules against families living there
Public housing unit	Many small in relation to numbers living there. Inadequate maintenance
Renting space to sleep at work	Usually total lack of facilities for washing/cooking. Lack of security
Renting a space to sleep in public buildings	Total lack of security/facilities. Payments to protection gangs or local officials
Building a house or shack in squatter settlement	Insecure tenure. Lack of public services. Dangerous locations etc.
Building a house or shack in an illegal subdivision	Sites are purchased and have degree of security. Some infrastructure. Often expensive
Building a house or shack on a legal land subdivision on the city periphery	Affordable plots on legal subdivisions often far from jobs
Invading empty houses or apartments or public buildings	Occupation illegal. Usually no services
Building a house or shack in government site and service or core housing scheme	Often far from jobs. Restrictions on employment activities from home. Eligibility criteria
Building a shack or house in a temporary camp	Often government's response to disaster situation. Infrastructure and services inadequate

Source: extracted from UNCHS (1996), but based originally on Hardoy and Satterthwaite (1989).

Plate 5.2 *Low-income housing*

a) Informal housing, Hanoi

Source: Huw Taylor, University of Brighton.

b) Public housing, Harare

Source: author.

Plate 5.2 *. . . continued*

c) Tenement blocks, Kolkata
Source: author.

The urban environmental challenge

The level of economic development and affluence is evidently a very important factor shaping the nature of environmental problems facing particular cities. As seen in Box 5.1, as cities get wealthier, industrial and energy-related pollution become more problematic, as does the inability to deal with wastes. These environmental burdens can be considered dispersed and delayed and will impact hardest in the future and for global communities. Clearly, there are many reasons, as discussed in Chapter 2, to consider these issues as immediate and profound challenges for the global community. However, these environmental burdens remain quite different to the immediate, local environmental health concerns characterising low-income slums. The nature of environmental problems in particular cities will also be influenced by the rate and scale of urbanisation itself and the degree of concentration of such growth. Fast-growing cities may provide particular challenges for planning and management but serious environmental problems can also occur in declining industrial centres and stagnant smaller towns, for example (UNCHS, 1996).

The geographical location of cities is a further factor shaping the nature of the environmental challenge: cities in cold climates consume greater levels of fossil fuels for domestic heating, for example. Mexico City is a widely cited case where altitude and topology (the city being surrounded on three sides by mountains) combine to present particular challenges for the dispersal of atmospheric pollutants, especially from industry and the motor car. Location is also a major determinant of the type and frequency of natural hazards that a city may experience. Eight of the ten most populous cities in the world are located on earthquake faults, for example, and are also at risk from a number of 'natural' disasters, as seen in Table 5.3. Coastal areas (where some of the highest rates of urban growth are currently occurring) are clearly more at risk from many of the impacts of climate change including sea level rise, tropical cyclones and flooding that are distinct from the environmental problems of cities further inland.

However, it is important to move beyond such broad patterns to understand the nature of the environmental burden of cities and how cities are responding to these. Wealth differences occur within cities (not just between them) and it is suggested that more competent,

Table 5.3 *Disaster risk in populous cities*

City	Population (million) in 2005	Earthquake	Volcano	Storms	Tornado	Flood	Storm surge
Tokyo	35.2	✓		✓	✓	✓	✓
Mexico City	19.4	✓	✓	✓			
New York	18.7	✓		✓			✓
São Paulo	18.3			✓		✓	
Mumbai	18.2	✓		✓		✓	✓
Delhi	15.0	✓		✓		✓	
Shanghai	14.5	✓		✓		✓	✓
Kolkata	14.3	✓		✓	✓	✓	✓

Source: adapted from UNHSP 2009.

accountable city governments can lessen environmental burdens across all levels of economic development (McGranahan and Satterthwaite, 2002). It is important not to underestimate the significance of these economic and political dimensions to sustainable urban development. In Figure 5.8 it was seen that many of the dimensions of poverty in urban areas relate to factors beyond income to include aspects of people's participation and voice within political systems and the protection of their rights in law, for example, and there are many external influences that shape local environmental concerns and the prospects for sustainable urban change in future. Indeed, Hardoy et al. (2001: 382) suggest that:

> it can even be misleading to refer to many of the most pressing environmental problems as "environmental" since they arise not from some particular shortage of an environmental resource (e.g. land or fresh water) but from economic or political factors that prevent poorer groups from obtaining them and from organising to defend them.

The following sections consider how these varied factors interact in specific locations at a range of scales towards further understanding the challenges of sustainable urban development in developing regions and how opportunities are being taken towards more sustainable patterns and processes.

The household and community level

In poor neighbourhoods of cities in the developing world, many of the most threatening environmental problems are found close to home. Regularly, poorer households use their homes as centres for income generation, homes also functioning as workshops, as stores for goods for sale, as shops or as bars or cafés. The environmental risks are often greater for women and children because of the longer hours spent at home and in the immediate vicinity. Women, for example, may combine, in the same space and time, piece-work for income with domestic duties such as child care. The environmental problems related to such activities in the home are diverse, but include the hazards to health associated with poor ventilation, inadequate light, the use of toxic or flammable chemicals and the lack of protective clothing. A high proportion of disablement and serious injury in cities of the developing world is caused by household accidents and these are strongly aligned to poor-quality, overcrowded conditions. Many low-income urban dwellings are constructed of flammable materials such as wood and cardboard and accidental fires are more common where families live in one room and where it is difficult to provide protection from open fires or kerosene heaters (UNHSP, 1996).

Indoor air pollution is also aggravated by the burning of low-quality fuels such as charcoal for domestic heating and lighting. The major impacts are on respiratory health, whereby irritant fumes cause respiratory tract inflammation, repeated exposure leading to the onset of chronic obstructive lung disease. Urban indoor pollution is estimated to be responsible for 3 million deaths every year (UNHSP, 2010). Young children are particularly vulnerable; they may be strapped to their mothers' backs during the course of daily indoor activities including cooking and suffer more as their smaller lungs are less able to cope with pollutants. In combination with malnutrition, smoke inhalation may further retard infant growth and raise susceptibility of children to other infections. It is suggested that acute respiratory infection may now be a bigger killer than diarrhoea amongst children (McGranahan and Murray, 2003).

One of the most critical determinants of human health (wherever people live) is access to adequate supplies of clean water. In 1980, the WHO estimated that 80 per cent of all sickness and disease worldwide was related to inadequate water (in terms of quantity and quality, and of solid and liquid wastes). Since then, where environmental improvements have been made in the quality of available water and in the disposal of excreta, illness and the burden of disease have been dramatically reduced and the impact on mortality has been even greater (World Bank, 1992). However, as many as 140 million urban residents currently have no ready access to safe water. Halving the proportion of people without access to safe water (by 2015 over 1990 levels) is a target within MDG7. On current trends, it is anticipated that this target will be met with 86 per cent of the population in developing regions having access to safe water supplies. However, the progress that has been made has largely been in rural areas (UN-DESA, 2010); drinking water coverage in urban areas (94 per cent in 2008) has remained almost unchanged since 1990. Table 5.4 illustrates the persistent challenge of extending improved urban drinking water and sanitation facilities in selected countries.

There are, however, acknowledged problems of data quality in such reporting. Many countries simply do not have data on who has 'safe'

Table 5.4 *Proportion of urban population with improved water sources and sanitation facilities, selected countries, 2008*

Country	% of urban population with access to improved drinking water	% of urban population with access to improved sanitation facilities
Angola	60	86
Argentina	98	91
Bangladesh	85	55
Brazil	99	87
Chad	67	23
Gambia	96	68
India	96	54
Indonesia	89	67
Mexico	96	90
Nigeria	75	36
Peru	90	81
Uganda	91	38

Source: compiled from UN Statistics Division www.unstats.un.org.

water (or sanitation that is 'adequate') in terms of how these significantly reduce health risks. There are also problems of outdated data that cannot account for the expansion of informal settlements or how existing facilities may fall into disrepair. There is a further tendency towards the overestimation of levels and 'adequacy' at the national level that often do not match with research within cities that reveals substantial local difficulties (see Satterthwaite, 2008b). Many governments, for example, classify adequate water provision as a tap within 100 metres of the house but in practice it is not uncommon for as many as five hundred people to share access to a single public standpipe (UNCHS, 2001). In many cases, the situation may be much worse, as seen in Table 5.5a. There are further questions concerning the length of time water is supplied each day when considering the adequacy of provision to ensure sufficient quantities of water at the household level or to secure the health benefits within a community (Table 5.5b). When access to water is restricted in these ways, for many low-income residents the option is either to draw water from surface sources (often, in effect, open sewers) or to purchase water (of unknown quality) from vendors. The costs of such water may be anything from four to one hundred times higher than the cost of water from a piped supply (WRI, 1996: 20). The inequities of this situation whereby poorer groups use less but have to pay more for

Plate 5.3 *Delivering basic urban needs*

a) Water in Jakarta, Indonesia
Source: author.

b) Fuel in Kairouan, Tunisia
Source: author.

Table 5.5 Questioning environmental improvement

(a) the adequacy of service

City	Numbers sharing public standpipe
Dakar, Senegal	1,513
Noukchott, Mauritania	2,500
Luanda, Angola	600–1,000

(b) the consistency of household supply (survey of 50 cities in Asia and Pacific in the mid-1990s officially stated as having 24-hour access to piped home supply)

City	Reported supply (hours per day)
Chennai	4
Karachi	4
Mumbai	5
Bandung	6
Kathmandu	6
Faisalabad	7

Source: Compiled from UNCHS (2001).

water was illustrated in Figure 2.6. In short, this further limits the amount of water used at the household level and is another important determinant of environmental health and individual well-being.

Whilst access to basic services like water and sanitation have a key role in poverty reduction and the improvement of living conditions within slums that are mandated under the MDGs, inequality evidently persists. Furthermore, there are concerns that a key policy response, of the widespread privatisation of basic utilities including water, sanitation and electricity in cities of the developing world, has put affordable access further from the most vulnerable groups in society. Whilst proponents of such economic liberalisation argue that the poor will benefit from properly financed utilities (and be saved reliance on water vendors, for example), this has been challenged by broad-based coalitions as seen in Box 5.3. In short, there is mounting evidence that privatisation leads to higher tariffs for basic services and fails to reach the poorest communities and those who have no access. Whilst the private sector can 'pick off' the most lucrative contracts and customers, the most 'hard to reach' people are left to public authorities whose capacities have been reduced by, for example, structural adjustment reform and the current global economic crisis. However, privatisation of basic service provision continues. Figure 5.11 identifies a number of concerns if the human right of all people to water is to be secured.

The inadequacy of urban water supplies serves substantially to explain the endemic nature of many debilitating and preventable diseases in cities of the developing world. Vulnerability to infection amongst low-income households is also enhanced by the inadequacies of urban facilities for the hygienic disposal of excreta or household

Box 5.3

The Cochabamba water wars, Bolivia

Cochabamba is a city in Bolivia with a population of over half a million, many of whom live typically on less than US$100 per month. In 2000, it was proposed (substantially promoted by the World Bank) to sell control of its public water system to a multinational consortium of private companies (Aguas del Tunari) with the prospect of water supply rates rising to US$20 per month. A new organisation, Coordinadora de Defensa del Agua (Coordinators in Defence of the Water), formed and a four-day strike shut down the city. When the government failed to respond, a public protest was called and a coalition of peasant unions, student groups and working-class unions as well as segments of the national security forces joined in the public protest against the privatisation. Riot police clashed with protestors and this resulted in more than 175 injured civilians as the public call for the cancellation of the contract rose. The support of a peasants' union that was fighting a parallel struggle against the privatisation of water provision in rural areas was gained. It organised road blocks extending into six of Bolivia's nine regions beyond the city. City residents stormed the local city hall where talks were being held, resulting in the arrest of 15 leaders of the Coordinadora. Subsequent demonstrations led to the release of the protest leaders, a reported cancellation of the contract and then denial on the part of the government and ultimately a declaration of a state of emergency, the suspension of rights to strike and legitimisation of the use of the army to prevent civilian unrest.

Rather than quell the discontent, the state of emergency fuelled further protests in Cochabamba and widespread discontent throughout the country to include university students, rural teachers and miners' unions, for example. Eventually, Aguas del Tunari itself withdrew from the deal and as it became clear that the privatisation of water provision in Cochabamba would not occur, Coordinadora called off the strikes. Whilst the immediate cause of the unrest has been removed, many residents of Cochabamba and more widely remain concerned about the growing pressures to privatise public resources in the country.

Source: compiled from UNCHS (2001).

garbage. Water and sanitation are intimately linked, not least where inadequate sanitation facilities lead to contamination of water and the spread of diseases such as cholera and diarrhoea that commonly occur in the cramped conditions of urban centres. However, it is a link that has also been overlooked. For people living in poverty, their need for drinking water and sufficient food, for example, may take precedence over concerns regarding poor sanitation. There is also a wider reluctance to 'talk about toilets' (Black, 2008) that extends to urban planners and across societies. The 'illegal' nature of slum settlements is also a factor that has limited government investments in sanitation and international donors have been steered towards the more overtly

Figure 5.11 *The key lessons of the recent privatisation of basic services*

....

● By its very nature, privatisation is increasingly forcing public authorities (both central and local) to become more profit-oriented in the provision of essential services. Among developing nations, where a significant proportion of the population lives in poverty, many segments of society are in no position to guarantee sufficient or adequate rates of return to the shareholders of private companies now providing basic services. Therefore, unless the rates charged by those utilities are subsidised in some way, already underprivileged people will likely be forced to forgo basic services altogether.

● Private corporate entities place strong emphasis on profit generation and cost recovery, which often has the effect of fragmenting service scope and delivery. If no potential or actual user can pay the full price for the new services, the project may become financially unsustainable.

● Private operators are accountable to investors rather than to the communities they serve. Of particular concern are the growing incidences of unethical practices by private suppliers and other institutions that aggressively push for increased privatisation. The need to strengthen participatory monitoring mechanisms could not be more acute, since privatisation is extremely difficult to reverse once effected.

Source: compiled from UNHSP (2010).

pressing poverty concerns and 'gaps' in provision in rural areas in comparison to urban areas.

The MDG target to halve the proportion of people globally without access to improved sanitation facilities is not expected to be met (UN-DESA, 2010). In 2008, 48 per cent of the population in developing regions (an estimated 2.6 billion people in both urban and rural contexts) continued to lack access to an improved sanitation facility. In urban centres, such access has only increased by 5 per cent over 1990 levels where an estimated 790 million people remain without improved sanitation. As with drinking water provision, the largest improvements in sanitation have been made in rural areas, although the gap between rural and urban coverage in some regions remains large, as was seen in Figure 4.4. The persistent challenge of improving urban sanitation for particular countries was shown in Table 5.4.

Whilst flushing toilets and underground systems of sewers are the most common means of disposal of human excreta in most cities of the developed world, they are rare in cities in developing regions and where they are present, they serve only a small proportion of residents. For households with no provision for individual or shared sanitation within their homes, 'open defecation' (OD) within ditches, streams, parks or other open spaces or into some container which is then thrown away ('wrap and throw') are often the only options. Provision of public or community toilets remains limited or of poor design and maintenance. Use may also involve a fee which people cannot afford. Examples of more sustainable sanitation provision are

Figure 5.12 *Proportion of population by sanitation practices 1980 and 2008 (percentage)*

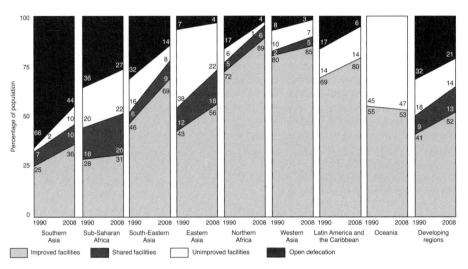

Note: Data for Latin America and the Caribbean and Oceania are not sufficient to provide regionally representative estimates of the proportion of the population who use shared sanitation facilities.

Source: UN-DESA (2010).

discussed below. Figure 5.12 identifies the progress that has been made at a regional scale towards enhancing access to improved sanitation. Whilst improvements in the extent of open defecation (the greatest threat to human health) are seen in all regions, the largest relative declines are in those regions where it was already practiced least (such as North Africa). In contrast, progress has been more limited in regions such as sub-Saharan Africa and Southern Asia where OD rates continue to be highest.

However, as with the data on drinking water, there are some problems with definitions and official reporting. There are also concerns as to how 'access' to sanitation is occurring in practice. Official statistics, for example, assume use of a facility that physically exists, yet research reveals many social, religious and cultural factors (as well as physical and economic) that shape access to particular sanitation facilities. There are particular problems for women and young girls, for example, in finding secluded spaces for open defecation. Research has also shown that many people may not use public or community facilities because they do not have the time to accompany their children and young girls to them or they are too far from their homes to use after dark (UNCHS, 2001). Adult women may feel too ashamed to use a public latrine in front of men during daylight hours and run the risk of rape when using such facilities after dark (Huggler, 2004). In particular cultures, there may be taboos regarding putting excreta underground where it will contaminate the dead, or layering one person's excreta on another (Black, 2008). Practical reasons for not using a facility may be because it is extremely foul smelling or is not constructed or maintained in a way that ensures personal dignity and social codes can be upheld. In short, all such poor sanitation contributes not only to individual poverty and ill-health, but also dangers for the wider community including direct exposure to faeces near homes and the contamination of drinking water. The cramped conditions of many informal settlements also aggravate the rapid transmission of disease between individuals, such as cholera, diarrhoea and tuberculosis.

The city and the wider region

Cities also present a number of environmental problems associated with the demands they place on ecosystems, resources and sinks of the surrounding region. Cities themselves involve the dramatic conversion of land use. Although on a global scale only around 1 per cent of the total land surface is under urban use (assessments vary according to and including the definition of 'urban' used), urban developments worldwide are encroaching on some of the last-remaining and most-valued reserves of natural vegetation, including mangrove swamps, protected wetlands, prime agricultural lands and forests. The expansion of built-up areas with city growth also involves transformation of the land surface, valley reshaping and the filling of swamps, for example, and the extraction and movement of rocks and other materials. In eastern Kolkata 4,000 hectares of inland lagoons have been filled to provide homes for middle-class families at the expense of the tenant families who formerly made a livelihood based on aquaculture in the region (WRI, 1996). In China, there were an estimated 80,000 'mass incidents' in rural areas in 2007 as local people protested against land and property seizures, corruption and pollution that was occurring as a result of urban expansion and rapid construction of housing and industry to serve urban interests (Buckley, 2010). The various solid, liquid and air-borne wastes generated within cities are also regularly transferred to the surrounding regions, with detrimental impacts on water bodies and land sites when liquid and solid wastes are disposed of without adequate treatment. Furthermore, demand from city-based activities for the products from forests, farmlands and watersheds located outside city boundaries present substantial challenges for sustainable urban development. Simply providing water for industrial and domestic uses is a problem for many cities. Mexico City, for example, depends on the Mexico Valley aquifer for 80 per cent of its water supply, and this has become so depleted that the city itself has sunk 10 metres since the 1930s as a result of excessive withdrawal of groundwater sources (UNHSP, 2010).

Although 'urban primacy' (i.e. the concentration of urbanisation in a few key cities within a country) may be less marked currently than in former times, there is still a tendency for industrial developments to be concentrated in a small number of urban centres. Regularly such developments are not subject to effective planning or pollution control. It was seen in Chapter 3 that the lack of stringent pollution control legislation generally in the developing world has been an important factor in attracting industrial production facilities (often transnational enterprises) to cities in these regions. Furthermore,

---

Box 5.4

The environmental impacts of Maquila developments on the Mexico–United States border

In the mid 1960s, the government of Mexico initiated a programme to promote industrialisation in the previously underdeveloped northern border region through creating EPZs incorporating 13 towns (Gwynne, 2008). Mexican and foreign factories were enabled to import machinery and inputs without paying tariffs on the condition that goods were re-exported. US companies were able to take advantage of cheap Mexican labour as well as US tariff regulations. The expansion of export-only assembly factories ('maquiladoras') substantially altered the distribution of population and urban development in Mexico. For example, the population of the 36 municipalities that adjoin the United States increased from 0.28 million in 1930 to 4 million by 1990 (UNCHS, 1996: 50).

By 1996, there were over 2,000 industrial plants employing more than 600,000 people. However, the expansion of employment has often been at substantial cost to the local environment: 'the fact is that Mexican border towns have become garbage dumps for millions of barrels of benzene solvents, pesticides, raw sewage and battery acid spewed out by foreign-owned maquiladoras' (Johnston-Hernandez, 1993: 10). The health impacts of such inadequate disposal of toxic wastes and chemical sludge are profound in urban developments where large proportions of the population depend on open water courses for their drinking water, for example. In the eastern border town of Matamoros, the rate of anencephaly (babies born without brains) was four times the national average, with tissues taken from the mothers of such babies showing the presence of pesticides, several of which have been banned in the United States (Johnston-Hernandez, 1993).

many countries have established export processing zones (EPZs) where manufacturing activities, often using transnational capital, are by definition encouraged and concentrated. In response to factors including the need to raise foreign exchange earnings to service debts and to open up their economies and attract foreign investment, EPZs have been a popular policy for governments of developing countries. TNCs are attracted to such zones by the tax exemptions and incentives offered and the typically low labour costs (Gwynne, 2008). Box 5.4 identifies a number of environmental challenges of industrial developments at the Mexican border with the United States (one of the largest concentrations of EPZs in the world).

Problems of air pollution have long been associated with cities, although there is currently much diversity worldwide in the relative importance of particular pollutants. In the developing world, sulphur dioxide and the concentration of suspended particulates are the major causes of urban air pollution resulting largely from industrial production and the burning of coal, oil and biomass fuels. In most cities of the more developed regions, tighter environmental regulations, measures to promote more efficient fuel use and the greater use of the least polluting fuels have reduced pollution from these 'traditional sources'. However, city-wide environmental problems also stem from activities other than industrial production. Congested roads and poorly maintained vehicles, for example, are a growing source of 'photochemical' (particularly lead and carbon monoxide) pollution as motor vehicle use per capita rises. Whilst it is difficult to ascertain the precise health impact of air pollutants, it is considered that air pollution is now eroding many of the previously gained health improvements in cities, particularly in Asia and Latin America (McGranahan and Murray, 2003). As many as half a million people died within Asia and the Pacific region in 2000 as a result of air pollution (UNEP, 2007). However, there are also examples of cities in the developing world leading international practice towards more sustainable transport that are proving key to health improvements for urban residents, ensuring efficiency of movement through the city that is essential to economic development and for lowering carbon emissions. One of the most heavily used yet lowest-cost (for the city authorities to provide and for passengers) public transport systems in the world is in Curitiba in Brazil. Here, over 2 million passengers daily use the Bus Rapid Transit system and no-one lives or works more than 400 metres from a bus stop (Simpson and Tuxworth, 2010).

In order to support both people and productive activities, cities depend substantially on inputs of raw materials and goods of various natures from the surrounding region. Whilst wealthy (powerful) cities have long had the capacity to draw resources from far beyond their immediate region, this capacity has greatly increased in recent decades, particularly as the relative cost of transportation has declined. Increasingly, food, fuel and material goods are drawn into cities from all over the nation and indeed the world (Potter et al., 2008). As already suggested, the larger and more prosperous cities tend to make greater demands, as consumption per head rises. Work on 'ecological footprinting' of cities (discussed further in Chapter 6) confirms that the environmental impact of more wealthy cities extends further into more distant surrounds than with lower-income cities, where the burden falls more locally. However, cities have many further indirect impacts on surrounding regions, through the commercialisation of land and agricultural markets for example. As discussed in Chapter 4, whilst such processes can be the basis for new economic opportunities (including through migration to urban centres), they are also often associated with environmental decline in rural areas through impacts on crops grown, the availability of labour at key points in the agricultural calendar, the economic viability of particular productive activities and even the expulsion of peasant farmers from their lands.

Clearly these types of process illustrate some of the limitations of the dualist distinction of 'urban' and 'rural'. This is confirmed, for example, in the case of urban demand for fuelwood and charcoal resources, where the supply necessarily comes from forested surrounding areas and, in many instances, pre-empts their use by rural residents; sources once available to rural inhabitants become unavailable to them as urban demand rises. This occurs through either deforestation per se or the commercialisation of fuelwood, which makes wood a commodity to be paid for rather than a resource to be collected from communal lands. Such regional environmental effects may be felt at considerable distance from the centre of demand (the city). For example, research has shown that fuelwood for the urban population of Bangalore comes typically from over 150 km away and from over 700 km in the case of Delhi (Hardoy et al., 2001). However, the consequences of such pressures for local people and local environments will also depend on the institutional context in these rural areas:

> urban charcoal demands are more likely to lead to deforestation if the producers are seeking out uncontrolled forests and have no intention of

returning, than if the producers are local residents harvesting wood from private or communal lands, having negotiated an agreement with the owners.

(McGranahan et al., 2004: 5)

Urban demand for charcoal and fuelwood can also serve to incentivise tree planting and environmental improvement.

Water is a resource with a strong regional dimension. Most cities rely on freshwater resources largely from within and around the urban centre as, unlike other resources, water cannot be imported easily from great distances. However, in most parts of the world, the spatial range of urban water withdrawals is expanding (McGranahan et al., 2004). Research in Africa suggests that whilst many major cities in the 1970s were using groundwater supplies as their primary sources, by the 1990s, their principal sources were more likely to be rivers more than 25 kilometres away (ibid.). Johannesburg in South Africa draws its water supply from 600 km away in neighbouring Lesotho (UNCHS, 2001). The construction of dams, canals and varied water diversion systems (including for hydroelectric power) regularly involves the loss of agricultural land and the displacement of people. The management of such 'cross-boundary' water resources presents substantial political and economic challenges, as identified in Chapter 2. It can also have major impacts on ecosystems and their functioning. Extraction of groundwater to serve Jakarta's water needs, for example, has led to saltwater intrusions extending 15 km inland (Hardoy et al., 2001).

The regional dimension of water is also illustrated in the environmental problems associated with the inadequate provision for the safe disposal and dispersal of industrial and domestic waste within cities. The city of Lima, Peru discharges 18,000 litres of untreated wastewater into the Pacific Ocean every second (UNHSP, 2006: 13) with global implications. Whilst the quantity of water moving into and out of urban areas is approximately the same, water is often returned to sources at far lower qualities than when supplied. As a consequence, water conditions downstream of urban centres tend to be more degraded than those upstream. As Hardoy et al. (2001: 109) suggest, many rivers running through cities in low- and middle-income countries are 'literally large open sewers'. They cite the case of the Yamuna River in India that is the source of drinking water for 57 million people including the population of Delhi. Yet 1,700 million litres of untreated sewage flow into that river each day from the city (as well as the industrial waste from 20 large,

25 medium and approximately 93,000 small-scale industries). The result is that the 500-kilometre section of the river below Delhi is virtually devoid of oxygen (eutrophicated) with serious implications for the contamination of domestic water sources, the undermining of agricultural production and the decline of fishing stocks.

In summary, the urban environmental challenge in the developing world is substantial in terms of extent and scope. Industrial developments and rising consumption in urban centres have been seen to be important factors in the degradation of urban environments, but it is also evident that for many low-income households, poverty and a lack of development closely define their core environmental challenges of daily living and working (i.e. the nature of the Brown Agenda). However, it has also been seen that some urban environmental problems diminish as cities become more productive and economically advanced, confirming that there are also opportunities that cities offer for more sustained development. In part as a result of the density of urban living, cities are places where a great variety of local initiatives and actions develop, often outside the formal or monetarised sectors, such as within citizens' groups, residents' associations and youth clubs, which are increasingly recognised to be essential for 'healthy' cities worldwide and a key resource for sustainable urban development actions.

Towards more sustainable urban development

It is evident from the sections above that reconciling the immediate and future needs of urban residents, managing environmental degradation and maintaining the economic advantages of urban development are substantial challenges for sustainable development in practice. In continuity with the progress shown towards sustainable rural development, sustainable urban development requires taking account of a complex set of natural, social and economic relationships, difficult political and economic trade-offs and a wide range of actors (Bartone et al., 1994). One of the most valuable resources available for sustainable urban development is now considered to be the capacity of citizens' groups to 'identify local problems and their causes, to organise and manage community-based initiatives and to monitor the effectiveness of external agencies working in their locality' (UNHSP, 1996: 427). However, the realisation of this capacity depends substantially on what happens at the city authority level, particularly in terms of the establishment of an effective system through which local people (including business interests) can participate in decision-making. In turn, city authorities remain responsible (despite increasing privatisation) for many functions which are critical to improving urban environments but are widely constrained in the developing world in part through the inadequate transfer of national finances to this level. There are also many issues, such as the reduction of greenhouse gas emissions or promoting more sustainable international trade practices and other essential elements of sustainable urban development, that require actions on behalf of institutions beyond the city level.

The following sections focus on two arenas that are considered key to meeting the sustainability challenges of urban centres and for the achievement of the Millennium Development Goals. In an era of rapid urban change and the diversity of cities and towns across developing regions, it is understood that taking decisions as close to urban citizens as possible is key (UNHSP, 2006, 2010). This suggests an important role for city authorities to devolve political, administrative and budgetary resources and functions to the local level but also to ensure participatory and transparent decision-making that improves the responsiveness of local policies and projects to citizens' priorities and needs. These form part of the challenges of 'good governance' discussed in Chapter 3. A second arena is the role of community groups and local innovation where new forms of governance that promote effective partnerships between local stakeholders, NGOs, private business, government and donors, for example, are needed if basic services and environmental improvements are to be provided for, by and with communities.

The effectiveness of city authorities

> Lessons drawn from 40 years' experience of national or international agencies demonstrate that most local problems need local institutions. Outside agencies, whether national ministries or international agencies, often misunderstand the nature of the problem and the range of options from which to choose the most appropriate solutions. They also fail to appreciate local resources and capacities.
>
> (Hardoy et al., 2001: 384–5)

City authorities worldwide are responsible for a range of urban management tasks including regulating building and land use, providing systems of water supply, sanitation and garbage collection, controlling pollution, managing traffic, delivering emergency services, and providing health care and education. They may not be directly responsible for all these tasks (the increased privatisation of service provision has been referred to above), but national and city authorities are responsible for providing the framework within which private as well as community-based developers operate, including the political context in which markets and local democracy work.

Issues of the capacity and responsiveness of local and sectoral institutions are an important determinant of the quality of the environment in a city. It was seen in Figure 5.8 that many aspects of urban poverty are linked to the limited capacity of local government agencies and departments to meet their responsibilities. Yet many city governments in the developing world are seriously constrained in terms of the finances and professional and technical competencies necessary to provide the investments, services and pollution control central to healthy urban environments. In many developing countries, city authorities depend on central governments for financial assistance to a much greater degree than in more developed countries. Furthermore, as identified in Chapter 1, governments themselves were often highly centralised, and often with authoritarian regimes, in many developing regions until quite recently. These governments often sought to consolidate their power through the establishment of (and the concentration of financial resources within) national urban development corporations and national housing authorities, for example. The result was often the construction of large, expensive infrastructural developments in urban centres, but inadequate financial resources at the local authority level to operate and maintain them.

However there have been major transformations in city governance across developing regions in the last two decades, including through processes of 'decentralisation' whereby power and responsibilities are increasingly being devolved from national to local governments. This includes decentralisation of administrative authority in decision-making and responsibilities, of political representation and of financial control over local budgets: 'Decentralisation exists in its most advanced form when elected local governments are empowered and capable of setting development priorities, making major development and expenditure decisions, and determining and collecting local revenue' (UNHSP, 2006: 170). These processes of decentralisation are complex and contested, particularly as overall government budgets have been cut (often under the pressure of debt servicing and the requirements of structural adjustment programmes) and are being restricted further by the current economic recession. A legacy of past failures of central government to transfer management responsibilities to local authorities has also been a lack of trained personnel at the local level. This is considered to be part of the relatively slow pace of decentralisation in many African countries (UNHSP, 2006). Decentralisation and wider 'good governance' also includes developing new partnerships for environmental management, which can be an intensely political task for example 'as different interests compete for the most advantageous locations, for the ownership or use of resources and waste sinks, and for publicly provided infrastructure and services' (Satterthwaite, 2008a: 264). It also includes understanding the needs, knowledge and capacities of the urban poor.

Decentralisaton policies are considered most advanced in Latin America. For example, the rise of democratic governments in many countries in the region in the late 1980s/early 1990s was also associated with progressive policies that have strengthened local government and enhanced the inclusion of grassroots movements and community organisations in planning decisions. Box 4.8 highlighted the role that the social movement Movimento do Trabalhadores Sem Terra has had in Brazil in influencing land and housing policy to better meet the needs of the poorest sectors of rural and urban society. Brazil was one of the first countries to introduce 'participatory budgeting' that allows community-led city councils to decide on health, education and other policies and how municipal budgets should be spent. It was the first country (in 1988) to include the 'right to the city' as part of its constitution towards challenging exclusionary urban developments that undermine human rights at the city level (UNHSP, 2010). Participatory budgeting is a flexible mechanism where each city may adopt different criteria for the selection of community representatives (who are generally from low-income districts) and for allocating budgets. It has resulted in the more active participation of civil society as well as more and better services to low-income groups (UNHSP, 2006), that is, supporting processes of more inclusive, poverty-focused and sustainable urban development.

Evidently, there is a key challenge for city authorities in finding new ways of working with other organisations at the local level. It was seen in the analysis of the nature of the environmental agenda at the household level, for example, that the extent of the problems and the shortfall in delivering environmental improvements are likely to remain beyond the capacity of local authorities alone to address. New partnerships are therefore essential to overcome this 'backlog' but may require significant changes amongst all involved institutions.

> The fact that capital is limited demands a more profound knowledge of the
> nature of environmental problems and their causes to allow limited
> resources to be used to best effect . . . potential solutions will need to be
> discussed locally and influenced by local citizens' own needs and priorities.
>
> (Hardoy et al., 2001: 398)

The analysis of the nature of the Brown Agenda above also highlighted
how the environmental concerns of the poor are intricately linked in
the same space and time to economic and social goals. However, the
traditional sectoral policies of urban authorities (and this applies in
more developed economies as well) may be ill-equipped to balance
such concerns. As Hardoy et al. (2001: 400) suggest:

> Most environmental problems are multidimensional, interconnected,
> interactive and dynamic which makes appropriate actions difficult for
> conventional government structures. The architects, planners and
> engineers who work for departments of housing or public works know
> very little about the environmental health problems faced by those they
> are meant to serve.

In continuity with the analysis presented in Chapter 4 in relation to
sustainable rural development, it is understood that there has often
been a mismatch between 'conventional professionally led urban
development strategies and the realities of urban development as
experienced by the poor' (Mitlin and Satterthwaite, 2004: 270). A
key response has been the development of interventions that have
sought to support strategies that emerge from people's own activities,
to better meet the needs of the urban poor and through establishing
new relationships between professionals/the state and local residents
that 'enable both parties to contribute to new solutions in urban
development' (ibid.: 271). However, there are substantial challenges
in understanding the very diversity of the urban poor and the
multiple economic, social and political factors that shape their
experiences in particular locations as considered above in terms of
sanitation needs and responses (see Joshi et al., 2011). The following
section considers some of the principles for more sustainable urban
developments that support community and local innovation.

Utilising the potential of community organisations and local innovation

There are plenty of examples of communities in urban centres of the
developing world over the last decades taking action to improve their

living conditions. Indeed, the total investment by individuals and groups in their homes and neighbourhoods has often greatly exceeded that made by city authorities (UNHSP, 2006). 'Often through no choice of their own, low-income households are *de facto* managers of the local environment' (WRI, 1996: 134). But it is only relatively recently that international institutions, aid agencies and national governments have recognised such initiatives as valuable. As identified above, in the 1960s and 1970s, for example, many national governments (with international backing) engaged in policies of squatter settlement destruction and removal. During this period, self-help housing, for example, was viewed with 'alarm and pessimism' (Potter and Lloyd-Evans, 1998: 144) and was seen as part of the problem of underdevelopment (thus necessitating clearance) rather than a reflection of poverty or even part of a solution. Just as with local rural development initiatives which are showing signs of sustainability, understanding is now emerging not only of the value of local initiatives per se, but of the preconditions which enable successful urban environmental management based on community organisations to be generated more widely.

In the late 1980s, five core lessons for successful sustainable urban development (Figure 5.13) were suggested on the basis of the work of NGOs and community organisations across 20 human settlement projects. They encapsulate often very wide-ranging changes in development (and research). In short, a principal prerequisite for sustainable urban development was to recognise that housing was not only a problem for central government, local authorities or the private sector but also a concern for communities themselves and '*given a chance*, poor communities hold the key to the solution of their own problems' (Conroy and Litvinoff, 1988: 252).

Figure 5.13 *Common characteristics of sustainable urban development*

1 Housing is also a people's problem
2 The need for building communities
3 The need for organising the community
4 The importance of outsiders
5 The importance of external funding

Source: Conroy and Litvinoff (1988).

In many cases, experience suggests that the nature of that 'chance' can be quite simple. For example, in a low-income district of Cali (Colombia's second largest city), an NGO, the Carvajal Foundation, assisted residents by building a warehouse in the middle of the 'squatter settlement' to provide space for manufacturers to sell construction materials directly to residents at wholesale prices (WRI, 1996). Until that time, a major factor in the inability of residents to build and improve houses had been the cost of construction materials which they had had to buy from retailers at some distance from the settlement. Once people had access to the construction materials, they were given further support such as in design and construction. Critically, the foundation played an important part in convincing the city authorities to approve the building plans and to set up a small office in the neighbourhood: 'Having preapproved building plans and easily obtainable permits was a valuable incentive for residents to build legal, affordable structures' (WRI, 1996: 138).

The experience in Cali confirms that securing support from government authorities was important in giving a sense of security in a community (essential for encouraging innovation) where the legal right to occupy land was still lacking. Experience of more sustainable housing projects also confirmed that a more holistic view of human settlements was required that was broader than the physical structure and included the critical issue of social organisation ('building the community' in Figure 5.13). Enabling communities to be stronger and better organised was essential for solving the immediate problems and for long-term benefits in the future. Just as with sustainable rural developments, community development in an urban context is something 'more than participation': it requires working with the poorest and most excluded groups, understanding and addressing their priorities in urban environmental management, as well as bringing together 'different voices' in the community.

One of the largest slum upgrading programmes in the developing world has been in been in Thailand where in 1992 the government, through its Urban Community Development Office (UCDO), supported community organisations with loans, small grants and technical support to form networks to negotiate collectively with city authorities. This has now been extended to involve the private sector, whereby UCDO has been merged with a Rural Development Fund to form a parastatal organisation. This organisation facilitates dialogue among communities in low-income settlements, and between these communities, the private sector, NGOs and the municipality. The Orangi Pilot Project (OPP) in Pakistan is widely cited as a successful and more sustainable example of planned, networked provision of sanitation for the urban poor (Box 5.5). It involves new partnerships between community groups, NGOs and public authorities. However, in both these cases, government recognition of informal settlements and tenure security on behalf of residents were essential in providing the enabling environment in which local residents could plan and invest and for developing the necessary partnerships with government authorities. Whilst formal research and monitoring of tenure security remains limited, it is suggested that between 30 and 50 per cent of urban residents in the developing world may lack any kind of legal document to confirm such security, that is, for the majority of urban inhabitants, their occupation of land and/or housing is 'either illegal, quasi-legal, tolerated or legitimized by customary or traditional laws, which can either be recognised or simply ignored by the authorities' (UNHSP, 2006: 95). This confirms the many challenges for sustainable urban development in future to provide solutions for the significant number of urban poor who continue to face the threat of eviction or indeed live on the streets (Joshi et al., 2011).

Box 5.5

The Orangi Pilot Project of low-cost sanitation in Pakistan

Orangi is a low-income settlement dating back to 1965 in Pakistan that now has over a million inhabitants. Most residents built their own houses without official help and there was no public provision for sanitation as the settlement developed. Whilst more affluent households constructed toilets connected to soakpits and some households close to creeks constructed sewerage lines emptying directly into these, public provision was absent and local enthusiasm to get local government agencies involved was low. However, most of the settlements within Orangi have been accepted by the government and land titles have been granted. In 1980, a local organisation was set up led by Dr Akhtar Hameed Khan, who was convinced that if local residents were fully involved, appropriate (and cheaper) sanitation systems could be developed. The OPP set up meetings in 'lanes' of adjacent houses, explained the benefits of improved sanitation, encouraged the organisation and election of local leaders and provided technical and planning assistance. Through the support of the OPP, local people in Orangi have built, financed and maintain a network of good-quality pour-flush toilets and a lane system of sewers that serve over 90,000 households (UNHSP, 2006). The costs per household were approximately one-fifth of what local utilities would have charged. The model has now been extended to most of Karachi's informal settlements. Significantly, as more local groups approached OPP and the scope of sewer construction grew, new arrangements were able to be made with government authorities. City authorities are now responsible for financing, managing and maintaining the main trunk sewers and waste treatment plants, and low-income urban residents do similarly in relation to the latrines, lane sewers and secondary sewers.

Experience of more successful sustainable urban development has also confirmed the importance of 'outsiders' for enabling communities to improve their own environments. Often it has been external NGOs which have been central in acting as support, advisory or action groups for community initiatives (or in securing these functions from other agencies on the community's behalf). Box 5.6 highlights the role of a regional NGO in India, the Society for the Promotion of Area Resources Centre (SPARC) and the ways in which it is working with other institutions including the World Bank, the municipal authorities and local community groups to foster improvements in sanitation and also in the wider living conditions of poorer groups. These examples confirm again that it is the relationship between community and local authorities that is critical for capturing the potential of community organisations and local innovation, and NGOs have often acted as a 'go-between' for residents and public authorities. A further characteristic is for NGOs to support and facilitate rather than take on what community organisations can do on their own, that is, working *with* rather than replacing existing institutions. In continuity with the experience in rural development, important factors in sustainability have often been the quality of the relationships between community groups and NGOs (as seen in Box 5.6). Interventions that are based on the

---

Box 5.6

The work of the Society for the Promotion of Area Resources Centre (SPARC) in Mumbai

If Indian cities are to grow in a healthy or sustainable way, we must learn how to partner and engage with informality and the urban poor. The Alliance of SPARC, the National Slum Dwellers Federation and Mahila Milan is currently working with 750,000 households across India.

(www.sparcindia.org)

Community toilets

In 2001, the Municipal Corporation of Greater Mumbai began a programme of slum sanitation improvement with support from the World Bank under its Mumbai Sewage Disposal Project. In contrast to previous practice whereby the city organised the construction of public toilets, paid contractors to build them and municipal departments maintained them, the programme sought to involve communities in the processes of design, construction, management and maintenance. The municipal authority successfully persuaded the World Bank to agree to change the funding and tendering arrangements to enable more community management and NGO involvement. On the basis of their previous successes in other cities, SPARC was contracted to provide toilets in 14 wards of Mumbai. Two hundred and eleven toilet blocks were developed in the first phase of the programme, benefitting over 200,000 people.

Local people were involved in the design and construction of the toilets, supported by engineers and architects from SPARC. There were significant differences in the design of the toilet blocks over conventional government models: they were bright and well ventilated, were better constructed allowing easier cleaning and maintenance, had large water storage tanks enabling water for bathing as well; each block had separate entrances for women and men giving women more privacy and saving time in queuing, and a block for children was included. Toilet blocks also included a room where a caretaker lived that meant that lower wages needed to be paid for maintenance. The cost of the toilet blocks was 5 per cent less than the municipal corporation's costing. SPARC is now working on the second phase of the project to build a further 150 toilet blocks. Through the success of the project, a Zero Open Defecation campaign is also being promoted across the country and a National Task Force for Sanitation was created in 2005.

Relocation and rehabilitation of railway dwellers

Since 2009, municipalities engaging in urban infrastructure projects in India have to make appropriate arrangements for the resettlement of households. SPARC (in association with other NGOs) is regularly appointed to work in a mediating and support role with the municipalities and the affected households towards more sustainable outcomes. There were many past examples of people being forcibly evicted by the police, then moved to areas at great distances from jobs and families and with insufficient infrastructure. In the late 1990s, the railway network in Mumbai was widely considered to be under severe need of repair and extension. However, informal settlement had encroached as close as five feet from the tracks in many locations and an estimated 24,000 families needed to be relocated if such improvements were to be made. To support affected families to engage in the planning and execution of the projects, the city authority sought the participation of SPARC and the National Slum Dwellers' Federation as well as a local community-based organisation, a women's saving cooperative comprising slum and pavement dwellers. The key role for SPARC was in initiating dialogue with families, assisting with moves, supporting the formation of societies of affected families, providing assistance with paperwork and helping to secure legal title to new housing. The government provided the land and the Railways provided the infrastructure. However, the communities demolished their own homes and were responsible for the new builds. To date, over 20,000 families have been successfully moved, voluntarily, to accommodation with secure tenure and simple amenities based on this community-led strategy. Using SPARC's experience across other housing projects, the costs of rebuilding are kept low by using family labour and community-based contractors and seeking out low-cost construction technologies. SPARC also supports people to access government subsidies and housing loans and lobbies the government, regulatory authorities and private financers for better housing policy and support.

Several factors underpin the success of the programme. The Mumbai Metropolitan Regional Development Authority in charge of the railway project was willing to give up some of the powers normally held by government agencies in such resettlement projects, giving responsibility to the NGOs for determining eligibility, obtaining information on the community and allocating housing, for example. While all these functions had previously offered opportunities for corruption and rent-seeking, the long-standing relationships between the community and the NGOs involved in the programme ensured levels of trust and good lines of communication. Households agreed the criteria for allocating accommodation in the new settlements and families formed lending cooperatives to assist families who had lost income as a result of the move. It is evident that the mutual trust and flexibility on the part of both community and government agencies were very much part of enabling poor people to act collectively for their own benefit and that of wider urban society.

Sources: Mitlin and Satterthwaite (2004); World Bank (2003a), www.sparcindia.org.

priorities defined by the urban poor themselves and where external agencies are engaged in a process of continuous support (rather than adopting a more piecemeal, project approach), lessons are learnt from the success of one initiative to build and stimulate further projects. As Mitlin and Satterthwaite (2004: 282) summarise, local NGOs recognise from the outset that poverty reduction requires more than an official recognition of the poor's needs:

> it has to include strengthening an accountable people's movement that is able to renegotiate the relationship between the urban poor and the state (and its political and bureaucratic apparatus at district, city and higher levels), and also between the urban poor and other stakeholders.

Conclusion

Whilst the Green Agenda has tended to dominate western environmental thinking and the actions of international institutions of development such as the World Bank and the United Nations, there is now a better understanding that it is the immediate adverse effects on survival for the urban poor of such basic processes as cooking, washing and working which ensure that the environmental challenges at the household level are of no less global proportions than global warming itself. Sustainable urban management in developing countries, as illustrated through this chapter, evidently requires interdependent actions across all levels of the hierarchy. If the actions of community groups are to be replicated widely to deliver environmental improvements on this scale, decentralised and democratic city and national governance is essential for enabling local groups to organise and for valuing their 'voice'. External assistance is required at all levels: in building capacity and competence amongst local authority planners and in fostering consensus and leadership within communities. All actions depend on innovative partnerships built on different approaches to understanding as well as new kinds of intervention.

In continuity with the lessons being learnt regarding progressing more sustainable rural development, urban development in the future must focus on the welfare needs of the poorest sectors of the towns and cities of the developing world. The urban environments of the poor are extremely hazardous to human health and the people themselves represent a substantial resource for the improvement of these environments. Enabling poor communities to take control of their own development can be a starting point for achieving levels of urban development and environmental change which are unlikely to be met by international and/or government finances. However, often this also involves safeguarding the needs of specific groups within poor communities against more powerful economic interest groups and addressing the continued challenges of insecurity of tenure for many households. Whilst the focus of this chapter has been the constraints on and conditions for sustainable urban development, the prospects for more sustainable urban development have also been seen to be tied closely to those of securing sustainable rural livelihoods.

It is also evident that the Brown environmental Agenda encompasses immediate and challenging issues that are explicitly linked to development concerns and the achievement of the MDGs. They are simultaneously about generating an income to live on day by day and the reality of resource degradation and danger. Unemployment is closely related to poverty and in turn to hazardous and deteriorating living and working conditions. The challenge of sustainable development includes a shift away from narrow sectoral programmes in urban development towards approaches which can both address and build upon these interdependent concerns.

Lessons are also being learnt (within the developed as much as the developing world) that a well-functioning urban system and an 'inclusive city' also depends on social stability, equity, integration and justice (UNHSP, 2010). Sustainable urban development requires new policies which reduce poverty and other forms of deprivation, but which are also socially and spatially inclusive. These are further factors lying behind the essential requirement for stronger local governance to ensure that context-specific and locally defined needs are addressed, that city authorities are more accountable to all citizens within their jurisdiction and that civil society is inspired to engage in participatory developments.

Summary

The rising concentration of the world's population in urban centres presents opportunities as well as challenges for sustainable development. Questions of urban sustainability are different from those in rural areas, but not unrelated.

There are many positive associations between rising urbanisation, the levels of human development and opportunities in sustainable development at a global scale.

The patterns and drivers of contemporary urbanisation in any particular country and point in time are complex and are often quite different to those that occurred in the past.

Slums are the most visibly obvious manifestation of the multiple deprivations of urban poverty.

The urban environmental challenges for many low-income residents are at the immediate household and community scales.

Processes of decentralisation of authority and decision-making from the state to city level and of privatisation of many basic services within cities of the developing world are two key features of contemporary urban development with key implications for sustainability.

The capacity, accountability and transparency of local authorities is a key factor in the prospects for sustainable development, but this also demands working in new ways with NGOs and citizen groups as well as the private sector.

Discussion questions

- Debate the advantages and disadvantages of the privatisation of water service provision in cities.

- Review the challenges of changing urban sanitation practices. To what extent are the barriers to change economic, social or political?

- Compile the evidence behind the suggestion that the challenges of the Brown Agenda are 'close to home' for the majority of residents in cities of the developing world.

- What are the environmental impacts of cities on surrounding rural regions? Think widely to include impacts on livelihood systems and relations between and within households in rural areas using your understanding from the previous chapter.

Further reading

Environment and Urbanization. An international journal that prioritises policy-relevant papers written in an accessible style.

Jones, G.A. (2010) 'The continuing debate about urban bias: the thesis, its critics, its influence and its implications for poverty-reduction strategies', *Progress in Development Studies*, 10,1 pp. 1–8. A good paper for understanding the continued relevance of the concept and outcomes of 'urban bias'.

Hardoy, J. et al. (2001) *Environmental Problems in an Urbanising World*, Earthscan, London. A landmark text in detailing the environmental problems of urbanisation at a range of scales.

UNHSP (United Nations Human Settlements Programme) (2010) *State of the World's Cities 2010/11, Bridging the Urban Divide*, Earthscan, London. The latest annual report of the UN body whose mission is to promote sustainable development. This report examines the drivers underpinning urban poverty and deprivation, the characteristics of the 'urban divide' and ways in which local authorities and national governments are working towards more sustainable urban development.

Websites

http://www.iied.org/human-settlements/home   A good site for investigating recent research and policy activities relating to human settlements. This group (part of the International Institute for Environment and Development) is committed to supporting NGOs and academics in the fields of poverty reduction, urban governance and more sustainable patterns of urban development.

www.unchs.org   The website for UN-HABITAT, the UN Human Settlement Programme that has a remit to promote sustainable development.

www.sparcindia.org   For further details of the work of the Society for the Promotion of Area Resources Centre (SPARC) that works in collaboration with other NGOs through many cities of India in the field of pro-poor and sustainable urban development.

http://www.un.org/millenniumgoals/   Gateway to understand more about the United Nation's work on and progress towards the MDGs.

6 Sustainable development in the developing world: an assessment

Learning outcomes

At the end of this chapter you should be able to:

- Understand the role of indicators in moving towards more sustainable processes and patterns of development
- Be familiar with a number of ways of assessing sustainable development
- Be aware of the strengths and limitations of indicators of and for sustainable development
- Reflect on the common but differentiated challenges of sustainable development into the future

Key concepts

Environmental assessment; ecological footprint analysis; green GDP accounting; decoupling.

Introduction

It has been seen through the preceding chapters of this book that the notion of sustainable development encompasses a wide range of concerns. These include the capacity of the planet to absorb the changes brought about by human activities and the substantially compromised development opportunities for many people in the world, particularly in the developing countries. From the investigation of the varied definitions and use of the term in Chapter 1, it was clear that there are different 'interests' in sustainable development and contested views of what should occur (the priorities for action and the nature of envisioned change) in future. As a result, the practice of sustainable development has to be understood as an inherently political and conflictual endeavour where those with more power are often best able to influence outcomes in their favour (Peet and Watts, 2004; O'Riordan, 2000).

The preceding chapters of this book have confirmed that the idea of sustainable development has been a strong influence in shaping many changes in environmental management and development worldwide: in terms of the way individuals act, businesses operate and communities organise themselves, for example, but also in directing the nature of state activities, in prompting the formation of new international institutions, and in fostering new ways in which all such organisations relate to each other in the search for patterns and processes of change which are more sustainable. One of the aims of the book was to highlight this 'institutional learning' and through considering the outcomes for people and environments in practice, to identify the continued challenges for further moves towards sustainable development. However, as more institutions of development declare an interest in 'sustainable development' as a policy goal and as further issues (such as global security) are articulated in terms of sustainable development, the requirement for continued critical questioning of the political nature of sustainable development becomes stronger. The experiences of more sustainable processes and outcomes (as in Chapters 4 and 5) confirm that sustainability rests on *inclusivity* and *reconciling* different needs and interests at the local level. Furthermore, it has been seen that new opportunities for sustainable development have emerged when previously dominant interests are challenged: as NGOs, for example, engage in international fora on environment and development and as rural development professionals work in more participatory ways that value local knowledges and priorities. Without continued critical reflection therefore, opportunities for more sustainable development are likely to be missed or compromised.

This chapter identifies the contribution of the expanding field of sustainable development indicators and appraisal towards assessing progress made. The final section reflects on evidence that has emerged through the substantive chapters of the book for a 'common future' for sustainable development.

Assessing progress towards sustainable development

In 1987, the WCED urged the development of new ways to measure and assess progress towards sustainable development. However, just as arriving at common definitions of sustainable development was seen in Chapter 1 as a substantial and ongoing challenge, this is demonstrated also in terms of the development of the means for assessment and measures of sustainable development. This field of assessment, appraisal and indicators within sustainable development is complex, with some key challenges as identified in Figure 6.1. For some authors, trying to 'tie down and measure' sustainability could be considered a 'futile exercise in measuring the immeasurable' (Bell and Morse, 2008: xvii) if sustainable development is understood not as a single thing, but a complex term open to a variety of interpretations and depending upon the various perceptions of the stakeholders in the specific problem context. Furthermore, the holistic nature of sustainable development (the many interdependent dimensions) and the importance of temporal scale (intergenerational issues) present substantial challenges for assessments and indicators which fundamentally seek to break down a complex system into its component parts. There are also concerns as to how to assess aspects of the process of moving towards more sustainable development, such as of participation or governance that have been seen through the text as central to sustainable development. Others acknowledge the problems involved, but accept that these have to:

> somehow be reconciled with the reality that everyone has a role to play in achieving sustainability – whether it is in policy development or consumption decisions. To make the right decisions we all need credible, accessible and timely information. Hence the advent of the indicator.
>
> (Chambers et al., 2000: 15)

Similarly, in publishing its strategy for sustainable development, the UK government stated that:

> Sustainable development is about ensuring a better quality of life for everyone, now and for generations to come. It means a more inclusive society in which the benefits of increased economic prosperity are widely shared, with less pollution and less wasteful use of natural resources. *To know whether we are meeting that goal, we need to be able to measure what is happening and monitor progress.*
>
> (DETR, 1999: 3 *emphasis added*)

Whilst a key driver of the development of assessment processes and indicators of sustainable development has certainly been to provide such information for decision-making and monitoring, there are also

Figure 6.1 *Key challenges in assessing sustainable development*

- Is sustainable development seen as a process or an end state? Is the assessment of the 'journey' or the 'destination'?
- Is the intention an assessment *of* sustainable development or *for* sustainable development?
- How can the more 'qualitative' aspects of sustainable development such as well-being or quality of life be measured? How can values such as justice or equity be measured?
- If sustainable development is broken down into a series of component measures within the environmental, social and economic spheres, for example, how can the synergies and trade-offs between them be captured?
- What is the appropriate time-scale over which to assess sustainable development?
- If there are many different 'interests' in and framings of sustainable development, who will decide what the outcomes should be and criteria within any assessment?
- If sustainable development is contextual and place specific, what purpose can indicators beyond the local level play?
- How important is the process of developing indicators itself (rather than the data generated) for contributing to sustainable development?

debates as to the purpose and use of assessment. For example, one view is that 'in general, people value what they measure' (World Bank, 2003b: 15). Another is that we should 'seek to measure what we value' (Blewitt, 2008: 175). For others, it is the processes of close reflection and dialogue that can occur as community groups, businesses and organisations, for example, engage in the development of indicators that enables the crystallisation and development of the very values that are required for sustainable development in the future (ESDinds Project, 2010).

In recent years in the context of what Talberth (2008) has termed the 'converging crises of the C21' of the realities of climate change, resource depletion and economic vulnerabilities, there has been substantial reflection as to whether the global community has the 'right' means and measures for assessing sustainability objectives. He suggests a consensus amongst governments, scientists and civil society on the need for a transition to an economic system where:

> progress is measured by improvements in well-being rather than by expansion of the scale and scope of market economic activity. We need to measure economic progress by how little we can consume and achieve a high quality of life rather than how fast we can add to the mountains of throwaway artefacts bursting the seams of landfills. We need to measure progress by how quickly we can build a renewable energy platform, meet basic human needs, discourage wasteful

consumption, and invest in rather than deplete natural and cultural capital . . . we need an economic system that is firmly ensconced within Earth's ecological limits . . . and is diverse, adaptable and resilient.

(p. 21)

The following sections consider progress towards meeting these challenges of measurement and assessment.

The use of indicators in sustainable development

The challenge of developing indicators of sustainable development has been taken up by a range of actors in civic, academic and development communities, at a variety of scales and for a number of purposes. In Chapter 3 it was seen how many businesses worldwide now report not just their economic performance in a particular year, but also in relation to additional social and environmental goals set for the company. There are many ways of conducting this so-called 'triple bottom line accounting' and as seen in Chapter 3 there are now networks such as the Global Reporting Initiative (GRI) that are supporting the development of consistent and quality systems for sustainability reporting globally. The GRI produce guidelines on the contents, quality and scope for sustainability reporting for organisations working across various sectors with the aim of promoting a more standardised approach and for sharing good practice. The development of quality sustainability indicators and reporting is recognised as an important means for both communicating the business commitment to sustainable development and for comparing organisational performance (Blewitt, 2008). The Millennium Development Goals can also be considered a set of indicators of sustainable development; the common, interdependent goals for development that are underpinned by specific, quantifiable and directly measurable targets relating to each goal. These are used to both direct and evaluate aid projects, for example.

Identifying indicators to predict and monitor the impacts of development interventions has also been a long-standing legal requirement in many countries. For example, the National Environmental Policy Act in the US has required developers since 1969 to consider the environmental implications of projects prior to their approval. Many multilateral and bilateral donor organisations now have some form of environmental assessment requirement, as seen in the case of the World Bank in Chapter 3, that includes

identifying measurable indicators for evaluation. The field of
'Strategic Environmental Assessment' (SEA) has also developed in
recent years towards assessing the impacts of development policies
and programmes (rather than discrete projects). Furthermore, it is
anticipated that these processes of assessment can (and should)
contribute to sustainable development. In 2001, the European
Community issued its directive on SEA within which one of the key
aims is to 'promote sustainable development (Article 1). Evaluation
of this field of environmental assessment and its contribution to
sustainability is beyond the scope of this text (but see Gibson, 2005;
Bond and Morrison-Saunders, 2011). However, much has been learnt
concerning both how to measure and account for environmental and
social impacts and what can be considered quality and effective
processes of assessment. This includes the importance of public and
stakeholder participation early and throughout the process. The public
right of access to information is central to sustainable development
and the basis of the Aarhus Convention signed by members of the UN
Economic Commission for Europe (ECE) of 1998:

> In order to contribute to the protection of the right of every person of
> present and future generations to live in an environment adequate to
> his or her health and well-being, each party shall guarantee the rights
> of access to information, public participation in decision-making, and
> access to justice in environmental matters in accordance with the
> provisions of this Convention.
>
> (Article 1)

Many governments, departments and local authorities are also
developing and using indicators to monitor their commitments
and achievements towards sustainable development. As seen in
Chapter 3, since the Earth Summit in 1992, many countries have
developed national and local strategies and action plans as part
of their commitment to Agenda 21 processes. The first sustainable
development strategy for the UK was published in 1994
(Department of Environment, 1994). The revised strategy in 1999,
entitled 'Quality of life counts', identified a number of priorities for
sustainable development, 15 'headline indicators' and over 150
measures of more specific issues and areas for action. Developing a
national set of indicators was considered important not only for
assessing the outcomes of policy initiatives (and thereby accounting
to their constituents) but also for educating people about what
sustainable development means and raising awareness of the further
actions that would be required. The multiple functions that these

Figure 6.2 *The intentions of the national core set of indicators*

- To describe whether we are achieving sustainable development overall
- To highlight key national-scale policy initiatives relevant to sustainable development and to monitor whether we are meeting key targets and commitments in those areas
- To educate the public about what sustainable development means
- To raise public and business awareness of particular actions which they need to take in order to achieve more sustainable development
- To report progress to international audiences
- To make transparent the trade-offs and synergies between sustainable development objectives

Source: DETR (1999: 16).

indicators were considered to perform are shown in Figure 6.2.

In 2005, a further revised strategy for the UK, 'Securing the future' was published. This took into account feedback from the Sustainable Development Commission (the UN body that oversees national strategies) that had identified over 20 key points on which the UK government was urged to take more decisive action. Development of the new strategy also included responses to a written consultation report, themed workshops organised by NGOs with an interest in sustainable development, a number of regional dialogue events and local community consultations run by trained facilitators. In turn a new indicator set was established on which the UK government reports annually. As seen in Figure 6.3 a number of measures, including those for individual well-being, were under development at the time. These have been reported since 2007 and confirm how sustainability indicators can change as what is valued (and considered important to measure) shifts, for example.

Figure 6.3 *UK government strategy indicators*

Indicator	Measure
1. Greenhouse gas emissions	Kyoto target and CO_2 emissions
2. CO_2 emissions by end user	Industry, domestic, transport (excluding international aviation), other
3. Aviation and shipping emissions	Greenhouse gases from UK-based international aviation and shipping fuel bunkers
4. Renewable electricity	Renewable electricity generated as a percentage of total electricity
5. Electricity generation	Electricity generated, CO_2, NOx and SO_2 emissions by electricity generators and GDP
6. Household energy use	Domestic CO_2 emissions and household final consumption expenditure

continued . . .

Figure 6.3 ... *continued*

Indicator	Measure
7. Road transport	CO_2, NOx, PM_{10} emissions and GDP
8. Private vehicles	CO_2 emissions and car/km and household final consumption expenditure
9. Road freight	CO_2 emissions and tonne/km, tonnes and GDP
10. Manufacturing sector	CO_2, NOx, SO_2, PM_{10} emissions and GVA
11. Service sector	CO_2, NOx emissions and GVA
12. Public sector	CO_2, NOx emissions and GVA
13. Resource use	Domestic material consumption and GDP
14. Energy supply	UK primary energy supply and gross inland energy consumption
15. Water resource use	Total abstractions from non-tidal surface and ground water sources and GDP
16. Domestic water consumption	Domestic water consumption per head
17. Water stress	(to be developed to monitor the impacts of water shortages)
18. Waste	Arising by (a) sector, (b) method of disposal
19. Household waste	(a) arising, (b) recycled or composted
20. Bird population	Bird population indices: (a) farmland birds, (b) woodland birds, (c) birds of coasts and estuaries, (d) wintering wetland birds
21. Biodiversity conservation	(a) priority species status, (b) priority habitat status
22. Agriculture sector	Fertiliser input, farmland bird population, ammonia and methane emissions and output
23. Farming and environmental stewardship	(to be developed to monitor progress in new stewardship schemes)
24. Land use	Area used for agriculture, woodland, water or river, urban (contextual indicator)
25. Land recycling	(a) new dwellings built on previously developed land or through conversions, (b) all new development on previously developed land
26. Dwelling density	Average density of new housing
27. Fish stocks	Fish stocks around the UK within sustainable limits
28. Ecological impacts of air pollution	Area of UK habitat sensitive to acidification and eutrophication with critical load exceedences
29. Emissions of air pollutants	SO_2, NOx, NH_3 and PM_{10} emissions and GDP
30. River quality	Rivers of good (a) biological, (b) chemical quality
31. Flooding	(to be developed to monitor sustainable approaches to ongoing flood management)
32. Economic output	Gross domestic product
33. Productivity	UK output per worker
34. Investment	(a) total investment, (b) social investment relative to GDP
35. Demography	Population of working age (contextual indicator)

continued . . .

Figure 6.3 ... *continued*

Indicator	Measure
36. Household and dwellings	Households, single person households and dwelling stock (contextual indicator)
37. Active community participation	Informal and formal, volunteering at least once a month
38. Crime	Crime survey and recorded crime for (a) vehicles, (b) domestic burglary, (c) violence
39. Fear of crime	(a) car theft, (b) burglary, (c) physical attack
40. Employment	People of working age in employment
41. Workless household	Population living in workless households (a) children, (b) working age
42. Economically inactive	People of working age who are economically inactive
43. Childhood poverty	Children in relative low-income households a) before housing costs, b) after housing costs
44. Young adults	16–19-year-olds not in employment, education or training
45. Pensioner poverty	Pensioners in relative low-income households a) before housing costs, b) after housing costs
46. Pension provision	Working-age people contributing to a non-state pension in at least three years out of the last four
47. Education	19-years-olds with level 2 qualifications and above
48. Sustainable development education	(to be developed to monitor the impact of formal learning on knowledge and awareness of sustainable development)
49. Health inequality	Infant mortality (by socio-economic group)
50. Healthy life expectancy	Healthy life expectancy (a) men, (b) women
51. Mortality rates	Death rates from (a) circulatory disease and (b) cancer, below 75 years and for areas with the worst health and deprivation indicators, and (c) suicides
52. Smoking	Prevalence of smoking (a) all adults, (b) routine and manual socio-economic groups
53. Childhood obesity	Prevalence of obesity in 2–10-year-olds
54. Diet	People consuming five or more portions of fruit and vegetables per day and in low-income households
55. Mobility	(a) number of trips per person by mode, (b) distance travelled per person per year by broad trip purpose
56. Getting to school	How children get to school
57. Accesibility	Access to key services
58. Road accidents	Number of adults and children killed or seriously injured
59. Social justice	(social measures to be developed)
60. Environmental equality	(environmental measures to be developed)
61. Air quality and health	(a) annual levels of particles and ozone, (b) days when air pollution is moderate or higher
62. Housing conditions	(a) social sector homes below the decent homes standard, (b) vulnerable households in the private sector in homes below the decent homes standard

continued ...

Figure 6.3 ... *continued*

Indicator	Measure
63. Households living in fuel poverty	(a) pensioners, (b) households with children, (c) disabled/long-term sick
64. Homelessness	(a) rough sleepers, (b) households in temporary accommodation (i) total, (ii) households with children
65. Local environment quality	(to be developed using information from the Local Environmental Quality Survey of England)
66. Satisfaction in local area	Households satisfied with the quality of the places in which they live (a) overall, (b) in deprived areas, (c) non-decent homes
67. UK international assistance	Net Official Development Assistance (a) per cent of Gross National Income (comparison with selected countries), (b) per capita (comparison with selected countries)
68. Well-being	(well-being measures to be developed)

The Bellagio STAMP (SusTainability Assessment and Measurement Principles) are a set of guidelines that have been established by international experts in the field of sustainability indicators and evaluation. They are intended for use by any group involved in assessing sustainable development: governments, international corporations or community organisations, for example. The guidelines refer to the assessment process as a whole, from establishing the vision of sustainable development through to ensuring that the continued capacity for assessing progress exists. Widespread participation and continuous adaptation are central to the processes, as seen in Figure 6.4.

Figure 6.4   *The Bellagio STAMP: SusTainability Assessment and Measurement Principles*

1. Guiding vision

Assessing progress towards sustainable development is guided by the goal to deliver well-being within the capacity of the biosphere to sustain it for future generations.

2. Essential considerations

Sustainability Assessments consider:

● The underlying social, economic and environmental system as a whole and the interactions among its components

● The adequacy of governance mechanisms

● Dynamics of current trends and drivers of change and their interactions

continued ...

Figure 6.4 *... continued*

....

● Risks, uncertainties and activities that can have an impact across boundaries
● Implications for decision-making, including trade-offs and synergies

3. Adequate scope
Sustainability Assessments adopt:
● Appropriate time horizon to capture both short- and long-term effects of current policy decisions and human activities
● Appropriate geographical scope ranging from local to global

4. Framework and indicators
Sustainability Assessments are based on:
● A conceptual framework that identifies the domains that core indicators have to cover
● The most recent and reliable data, projections and models to infer trends and build scenarios
● Standardised measurement methods, whenever possible in the interest of comparability
● Comparison of indicator values with targets and benchmarks, where possible

5. Transparency
The assessment of progress towards sustainable development:
● Ensures the data, indicators and results of the assessment are accessible to the public
● Explains the choices, assumptions and uncertainties determining the results of the assessment
● Discloses data sources and methods
● Discloses all sources of funding and potential conflicts of interest

6. Effective communication
In the interest of effective communication, to attract the broadest possible audience and to minimise the risk of misuse, Sustainability Assessments:
● Use clear and plain language
● Present information in a fair and objective way, which helps to build trust
● Use innovative visual tools and graphics to aid interpretation and tell a story
● Make data available in as much detail as reliable and practical

7. Broad participation
To strengthen their legitimacy and relevance, sustainability assessments should:
● Find appropriate ways to reflect the views of the public, while providing active leadership
● Engage early on with users of the assessment so that it best fits their needs

8. Continuity and capacity
Assessment of progress towards sustainable development requires:
● Repeated measurement
● Responsiveness to change
● Investment to develop and maintain adequate capacity
● Continuous learning and improvement

....

The development of indicators in practice

One of the best known practical methods for exploring the human impacts on environment (and how far society at various scales may be away from sustainability targets) has been the 'ecological footprint' (EF). It was originally developed in the early 1990s and is a complex and rapidly expanding methodology. In short, the ecological footprinting is a quantitative assessment of all the biophysical resources needed to support the consumption of particular groups of people, a country or city, for example, in terms of the raw materials and energy used to extract, produce and transport manufactured goods and for their disposal. It is typically expressed in terms of hectares of biologically productive area (of world average productivity) that are required to support that activity. EF is clearly related to the notion of carrying capacity, but rather than asking 'how many people can the earth support?', ecological footprint analysis asks 'how much land do people require to support themselves?' (Chambers et al., 2000: 59).

Most commonly footprint analysis has been calculated and expressed for countries, regions and cities, but can also be applied to products where the 'Environmental Rucksack' is the total amount of energy and raw materials required to extract, transport and manufacture a particular good. All such analyses rest on the fundamental premise that it is possible to estimate with reasonable accuracy the resources consumed and waste generated and in turn whether these can be converted to an equivalent biologically productive area necessary to support those functions (Wackernagel, 1998). The precise methods and calculations used will depend on the level of detail required, the data available and the purpose of the assessment. The Global Footprint Network was founded in 2003 to develop, coordinate and standardise the accounting methods used. It has over 50 partner organisations from public, business and civil society sectors. Ecological footprinting is not without its critics, but is generally accepted as 'a great step forward' (Levett, 1998: 67) particularly for capturing ecological and spatial/distributional aspects of sustainability (it is less good on economic aspects) and in assisting the visualisation of human impacts on the earth. Its 'simplicity' is proving to be valuable in assisting public understanding and in presenting a starting point for debate towards better planning.

Carbon footprinting is an expanding (and complex) field that can be considered a form of EF. As seen in previous chapters, the challenge of moving towards lower carbon futures has prompted new ways of assessing humanity's 'carbon footprint' on the Earth's atmosphere. These include assessing and monitoring the tons of carbon emitted by whole countries (as in reporting under the Kyoto Protocol), the manufacture and transport of particular products (the basis of carbon-labelling schemes) and the carbon intensity of economic output (emissions per unit of GDP). A different method of carbon footprinting involves quantifying the area of the Earth's surface needed to sequester those emissions. However, all such indicators are complex to measure in practice, particularly when taking account of the emissions that are associated with the energy and materials used within traded goods. There are substantial difficulties also in measuring carbon sequestration under different land uses and forestry types, for example, that continue to challenge international agreements on responses to climate change (as well as particular initiatives such as REDD+ seen in Box 3.3). Like EF more widely, carbon footprinting is proving a useful communication tool and way to monitor progress as well as to advance and assess such policy measures.

Assessment of ecological footprints is also a central part of the Happy Planet Index reported for countries worldwide since 2006 by the New Economics Foundation. This index takes the conventional approach to EF of quantifying the amount of resources used further to consider the 'efficiency' with which such ecological inputs are then converted into human development outcomes in terms of life expectancy and life satisfaction. The Index uses existing data on life expectancy as collected by the UNDP within its annual *Human Development Reports* and on life satisfaction largely from existing surveys undertaken by the pollsters Gallup in their World Poll (which now extends to 112 countries). Life satisfaction is multiplied by average life expectancy and divided by the ecological footprint (using data from the World Wildlife Fund's *Living Planet Report*) to produce the HPI. In 2009, Costa Rica came top of the HPI league, as seen in Table 6.1. It was followed by nine countries also in Latin America. The highest ranking western nation was the Netherlands (ranked forty-third). The index highlights that it is possible to live long happy lives with a much smaller ecological footprint than found in the highest-consuming nations. This is confirmed in the case of Costa Rica, which has a footprint less than a quarter the size of the USA but very similar life expectancy and life satisfaction. A key finding was that in general, countries classified by the UN as

Table 6.1   The Happy Planet Index (selected countries)

HPI ranking	Country	Life expectancy (years)	Life satisfaction	Ecological footprint (global hectares)	HPI
1	Costa Rica	78.5	8.5	2.3	76.1
2	Dominican Republic	71.5	7.6	1.5	71.8
9	Brazil	71.7	7.6	2.4	61.0
16	Indonesia	69.7	5.7	0.9	58.9
20	China	72.5	6.7	2.1	57.1
23	Mexico	75.6	7.7	3.4	55.6
35	India	63.7	5.5	0.9	53.0
43	Netherlands	79.2	7.7	4.4	50.6
51	Germany	79.1	7.2	4.2	48.1
74	UK	79.0	7.4	5.3	43.3
114	USA	77.9	7.9	9.4	30.7
118	South Africa	50.8	5.0	2.1	29.7

Source: compiled from NEF (2009a).

'medium human development' fared better than those in 'low' or 'high' development categories, suggesting that human well-being does not depend on high levels of resource consumption.

In recent years, there has also been mounting interest in the development of measures that adjust conventional indicators of economic development to better reflect the elements that matter for sustainable development. These are collectively termed 'green GDP' accounting. As seen in Chapter 2, there has been long-standing dissatisfaction within development studies regarding Gross Domestic Product (and Gross National Product) for measuring development, poverty and human well-being. There are also concerns for the way GDP fails to account for the environmental costs associated with development, as highlighted in Figure 6.5. Some kind of green accounting programme is now in place in over 50 countries and are planned for a further 20 (Talberth, 2008). The focus is on 'adjusting' GDP to account for the costs of resource depletion and damage and/or 'augmenting' it such as through adding the economic value of higher education and voluntary work. All aim to provide a better measure of sustainable economic welfare rather than economic activity per se.

The Genuine Progress Indicator (GPI) is one of the most established examples of an adjusted or green GDP measure. The GPI

Figure 6.5 *The failings of Gross Domestic Product as a measure of progress*

It is beyond dispute that GDP fails as a true measure of societal welfare. While it measures the economic value of consumption, GDP says nothing about overall quality of life ... GDP gives no indication of sustainability because it fails to account for depletion of either human or natural capital. It is oblivious to the extinction of local economic systems and knowledge to disappearing forests, wetlands, or farmland; to the depletion of oil, minerals, or groundwater; to the deaths displacements, and destruction caused by war and natural disasters. And it fails to register costs of pollution and the non-market benefits associated with volunteer work, parenting, and ecosystem services provided by nature. GDP is also flawed because it counts war spending as improving welfare even though theoretically, at best, all such spending really does is keep existing welfare from deteriorating.

(Talberth, 2008: 19)

GDP is ... merely a gross tally of products and services bought and sold, with no distinctions between transactions that add to well-being, and those that diminish it. GDP, rather than leading us down the right path, points us in a completely random direction. It is no measure of progress. It increases with polluting activities and then again with their clean-up.

(www.foe.co.uk/community/tools/isew/annex1.html, accessed 13/7/10)

differentiates between economic activity that diminishes natural and social capital and activity that enhances these capitals. It uses the same personal consumption data as GDP but makes deductions for income inequality, crime and environmental degradation and loss of leisure, for example. It adds factors relating to the value of public infrastructure and of housework and parenting, volunteer work and higher education. The indicators of environmental decline used in the GPI include the costs of air, water and noise pollution, loss of wetlands, farmlands and primary forest, depletion of non-renewable energy resources, carbon dioxide emissions damage and the costs of ozone depletion. A key problem of such measures is the difficulty of putting a monetary value on elements such as voluntary activity and wetland loss. By definition, voluntary activity is not paid for and wetland resources (and the ecosystem services they provide such as supporting wildlife or recreation) are not currently part of any formal market system. However, the GPI and related measures have been an important part of thinking through not only how to measure a country's progress in more meaningful terms (i.e. to reflect what is now valued, including ecological health and life satisfaction) but also for questioning more broadly the model of economic development that underpins these outcomes (Box 6.1).

The recent global economic crisis has also added urgency to this endeavour. A number of international organisations are now taking on the challenge of how to assess progress towards visions of future

Box 6.1

Questioning the routes to sustainable development: what indicators can reveal

There is no doubt that higher incomes are central to an improved quality of life. However, there is mounting evidence that increases in 'conventional economic growth', that is, GDP, does not couple well with actual human well-being. For example, the Happy Planet Index has shown that some of the poorest countries of the world (in sub-Saharan Africa) as defined by low GDP also had the lowest HPI scores. However, 'middle' income countries scored well on the HPI and there was evidence that further increases in GDP per capita are associated with lower scores. Table 6.1 showed that the US scored only marginally better than South Africa despite a GDP per capita that is almost seven times as large. Furthermore, calculations of the Genuine Progress Indicator for the US reveal that it was at its highest in 1975 when GDP per capita for the country was approximately half of what it is today (Redefining Progress, 2006). This suggests that the costs of economic growth have significantly outweighed the benefits since that time.

These findings confirm that there are levels of GDP/income above which well-being stops increasing and becomes subject to 'diminishing returns', that is, the efficiency of achieving well-being decreases dramatically and many countries are now in a period of 'uneconomic growth'. The data from both the HPI and the GPI suggests that in many western societies, GDP could be lowered and fewer resources consumed with little negative impact on well-being. This 'de-coupling' of well-being from further economic growth as conventionally measured not only questions if the global community has the right measures but whether further increases in GDP per capita in particular areas of the world are actually desirable (Jackson, 2009).

development that are built around sustainability and well-being. For example, there are ongoing projects hosted by the Organisation of Economic Cooperation and Development (OECD) on *Measuring the Progress of Society* and by the European Commission on *GDP and Beyond.* The 'Stiglitz Report' has received widespread international attention. In 2008, the French president (Nicolas Sarkozy) appointed Joseph Stiglitz as chair and Amartya Sen as advisor (both Nobel prize winners in economic sciences and professors at Columbia and Harvard Universities respectively) to a commission to assess the strengths and feasibility of the varied alternatives to GDP available and to identify the additional information needed to produce more relevant indicators of social progress. President Sarkozy was concerned as to the existing state of statistical information about the economy and society but also how the economic crisis at the time had taken so many governments by surprise. The commission reported failings in both existing measurements of progress and how governments (and other market participants) interpreted them:

> We are now living one of the worst financial, economic and social crises in post-war history . . . it is perhaps going too far to hope that had we had a better measurement system, one that would have signalled problems ahead, so governments might have taken early measures to avoid or at least to mitigate the present turmoil. But, perhaps had there been more awareness of the limitations of standard metrics, like GDP, there would have been less euphoria over economic performance in the years prior to the crisis; metrics which incorporated assessments of sustainability (e.g. increasing indebtedness) would have provided a more cautious view of economic performance.
>
> (Stiglitz et al., 2009: 8–9)

These reviews are now informing much country-specific work. The UK government, for example, has committed to work to develop a set of indicators that could be used across all government policies, programmes and projects that improve understanding of the relationship between specific interventions and the well-being of individuals, communities and the environment. Across many different policy areas (i.e. not solely economic development), it is recognised that more informed public policy (that can also turn out to be more cost effective) depends on valuing the outcomes that matter to individuals in terms of how their lives, their communities and the environment change as a result of policy (NEF, 2009b). This includes measuring the more subjective dimensions of change that can be hard to quantify. Developing measures to assess 'happiness' is now part of the work of the UK Government Office for National Statistics for example (Thomas and Evans, 2010). It also rests on looking closely at how different indicators are linked to each other to provide a more holistic view of change that is central to sustainable development. That integrated view is also key to understanding and balancing the tradeoffs that are inevitably required in policy-making; better evidence and understanding of linkages enables the choices being made within development interventions to be made explicit and transparent to the constituencies involved that are also key to more sustainable policy and practice.

For many organisations involved in working towards more sustainable development, particularly civil society organisations (including environmental groups, faith-based groups, charities and social enterprises), their work is explicitly what is termed 'value-driven', that is, it is precisely the achievement of the more 'intangible' outcomes such as empowerment, social cohesion or respect for the environment that underpins their activities. As identified in Chapter 1, there are now diverse initiatives that consider that it is a fundamental change of values (in 'hearts and minds') that is the central to the challenge of sustainable development. The Earth Charter, for example, was the outcome of global consultation processes and is a vision of sustainable development built on values. It is endorsed by hundreds of organisations including international NGOs, national governments, international conservation and humanitarian organisations and national environmental networks. The shared values and principles of the Earth Charter that these organisations are committed to further through their work are shown in Figure 6.6. Research into developing values-based indicators that support such organisations to examine, measure and monitor values within their work is expanding (see ESDinds Project, 2010; Dahl, 2012).

The strengths and weaknesses of indicators

The work on developing indicators of sustainability raises the issue of sustainable development as a goal or a process of change and whether the practice or endeavour of developing indicators is itself an assessment *of* sustainable development or *for* sustainable development. The MDGs and the UK government indicators, for example, essentially seek to define a vision or condition of sustainability in terms of a set of economic, social and environmental criteria. They highlight what is being valued/sought and set out how it will be measured through a series of indicators *of* sustainable development. A strength of such sustainability indicators were seen to include the flexibility to add and refine indicators, particularly as new data sources become available, as new ways of measuring some of the more subjective outcomes are developed and also as what is being valued by society evolves. Critics continue to point to the essentially reductionist nature of the endeavour, questioning how the components of sustainable development in practice can be assessed by a series of measures of inputs and outputs, how the complex interlinkages and possible tradeoffs between indicators can be captured and question the very idea of sustainable development as an end point rather than a process of change. Evidently, great care is needed when understanding and interpreting any particular set of

Figure 6.6 *The shared ethical framework of the Earth Charter Initiative*

The interdependent principles for a sustainable way of life:

I. Respect and care for the community of life

    1. Respect Earth and life in all its diversity

    2. Care for the community of life with understanding, compassion and love

    3. Build democratic societies that are just, participatory, sustainable and peaceful

    4. Secure Earth's bounty and beauty for the present and future generations

II. Ecological integrity

    5. Protect and restore the integrity of the Earth's ecological systems, with special concern for biological diversity and the natural processes that sustain life

    6. Prevent harm as the best method of environmental protection and, when knowledge is limited, apply a precautionary approach

    7. Adopt patterns of production, consumption and reproduction that safeguard Earth's regenerative capacities, human rights and community well-being

    8. Advance the study of ecological sustainability and promote the open exchange and wide application of the knowledge acquired

III. Social and economic justice

    9. Eradicate poverty as an ethical, social and environmental imperative

    10. Ensure that economic activities and institutions at all levels promote human development in an equitable and sustainable manner

    11. Affirm gender equality and equity as prerequisites to sustainable development and ensure universal access to education, healthcare and economic opportunity

    12. Uphold the right of all, without discrimination, to a natural and social environmental supportive of human dignity, bodily health and spiritual well-being, with special attention to the rights of indigenous peoples and minorities

IV. Democracy, nonviolence and peace

    13. Strengthen democratic institutions at all levels, and provide transparency and accountability in governance, inclusive participation in decision-making and access to justice

    14. Integrate into formal education and life-long learning the knowledge, values and skills needed for a sustainable way of life

    15. Treat all living beings with respect and consideration

    16. Promote a culture of tolerance, non-violence and peace

indices of achievements as measured. Of particular importance are the processes through which they are identified, how different interests and stakeholders are involved and the debate and dialogue that underpins their use and continued development, as identified in the Bellagio principles (Figure 6.4).

Simple headline indicators of sustainable development as in the UK strategy (and of 'un'sustainable development as in ecological footprinting) can also be an important way to consider complex issues in a reasonably integrated or holistic way. They are also useful in considering overall trajectories through annual assessments over time and for showing 'how far away' from sustainable development particular places or groups are at a specific point in time. Where the process of developing indicators includes broad participation and consultation, this not only informs but also educates stakeholders. In this way, it is less important that the indicators are 'ideal', but the process of their development is important for encouraging ownership and debate, that is, the role of developing indicators is *for* sustainable development.

There is also evidence that the substantial work of independent think tanks (such as the New Economics Foundation) and NGOs (such as Friends of the Earth and Redefining Progress) in developing and communicating alternative measures of 'progress' are now influencing governments. As seen, indicators such as the HPI and GPI seek to measure the components of change that work to undermine or support the outcomes and activities that matter to people, including ecological decline and life satisfaction. The evidence that these measures have provided concerning, for example, how what is valued in society has become increasingly decoupled from economic growth, is proving important for governments to rethink not only how they measure the outcomes of policy but also how 'development' to date has been achieved.

Fundamentally, as has been highlighted throughout the text, sustainability is a complex and contested notion with various interpretations and differentiated interests in the nature of future change. Any particular set of indicators or measurements of achievement therefore need to be understood and interpreted in this context. However, it has also been seen in a range of arenas (particularly within Chapter 3) that access to information, greater transparency and accountability, for example, have been powerful forces for change. It has also been seen how a lack of democracy, together with social conflict and environmental injustice, all threaten resources and jeopardise intragenerational well-being. The continued development of indicators, particularly in the social spheres of sustainability and towards capturing these less tangible (including values-based) dimensions, will be an important part of further action towards sustainability. Additional work is also needed (as identified in Chapter 4) in enhancing participatory methods for monitoring and evaluating all kinds of policy and projects, that is, assessing change through processes that include people being affected by impacts or affecting those outcomes. Only in this way will it be understood whether projects are making a difference that matters to the people who are living with those changes.

A common future?

Almost 25 years ago, in presenting the report of the World Commission on Environment and Development, the Chair, Gro Harlem Brundtland, stated that 'we live in an era in the history of nations when there is a greater need than ever for co-ordinated political action and responsibility' (WCED, 1987: x). The challenges of sustainable development were identified as 'cutting across the divides of national sovereignty, of limited strategies for economic gain, and of separated disciplines of science' (p. x). The call was 'for a common endeavour and for new norms of behaviour at all levels and in the interests of all' (p. xiv).

These same pleas are currently being voiced as the global community attempts to respond to the unfolding economic crisis. Coordinated political action and responsibility is currently being negotiated by finance ministers within the Euro zone to determine the economic future of whole countries such as Greece, and avoiding a return to protectionism is considered key to the economic future of the European Union as a region. The 'Occupy movement', a coalition of campaigners whose actions have created attention in many cities across the world, has highlighted the extremes and injustices of global capitalism and the new norms of behaviour that are required particularly within the global banking and finance industries but also within governments to ensure these institutions operate in the interests of all. Evidently, the challenges of reconciling different interests and ensuring inclusivity even as common futures are acknowledged continues to prove a difficult and ever more urgent challenge. However, it is also suggested that:

> We are now at an exciting stage in the awkward but vital, transition to sustainability. In international government, in national strategies, in business, in community action and in individual behaviour and outlook, we are beginning to witness a dawning realisation that global humanity has to shift if future generations are to survive with any meaningful sense of prosperity and well-being.
>
> (O'Riordan, 2009: 307)

Whilst the emphasis within this text was has been on the particular environment and development challenges and progress in the developing world, the interdependence of different regions of the world has been demonstrated repeatedly: where change in any one element of the world's atmospheric system has implications for the functioning of the system as a whole (and all organisms living within it); where a substantially globalised economic system operates through geographical differences (of both the physical and human environments); and where individual human rights are now understood to connect to global security and to the global challenge of sustainable development.

Throughout the text, the detail of the patterns and processes of development (and the outcomes of policies and projects towards sustainable development) has shown substantial evidence of the global nature of the challenges of sustainable development. Climate warming presents the archetypal global environmental problem and challenges the social sciences as well as the earth sciences, international relations and all countries to find lower-carbon processes and patterns of production and consumption. Worldwide there are also local problems of natural resources decline, pollution and degradation of the varied services that ecosystems provide and that are essential to human well-being, as revealed within the Millennium Ecosystem Assessment. The expansion of Multilateral Environmental Agreements, the outcomes of core summits on environment and development and the international commitments to the Millennium Development Goals have been provided as evidence that the common nature of the global challenge of sustainable development has been recognised at the highest political level.

The economic recession that started in the US within the financial sector in 2008 but quickly moved to affect the productive sectors in countries across the world has confirmed the common experience of the volatile and unstable nature of economic development globally. It has also led to deep questioning of how future growth can be secured. Evidence of the considerable interest in the 'win-win' opportunities of stimulating new patterns of economic growth that could also lay the basis for lower-carbon growth trajectories in the future was seen in Chapter 3. This was evident in the 'merging' of climate and development agendas within reports of key institutions in development, in country and regional commitments regarding the use of renewable energy technologies and in government support for 'green investments' as part of their overall stimulus spending towards assisting economic recovery.

Concerns for the justice encompassed in moving towards sustainable development, for how environmental degradation (and environmental improvements and the impacts of environmental policy decisions) are distributed across society (and spatially) have also been seen to be understood as challenges within more economically advanced countries as well as within developing regions. The environmental justice movement in the US centres on urban injustices of pollution, hazardous waste and environmental dangers that are spatially concentrated in poor and minority neighbourhoods.

'Environmentalism of the poor' extends the focus to the many thousands of local ecological distribution conflicts and injustices, urban and rural, that impact on the livelihoods of the poorest groups within countries, that are often a key cause of poverty and further social conflict and that are explained in terms of the growing requirements for natural resources and consumption goods by more wealthy groups and countries. The common, shared concern globally about the injustices of patterns and processes of development and understanding of the interlinked challenges in environment, economy and societies is encapsulated in the rising prominence of social movements, seen throughout the text. Whereas in the past, international non-governmental organisations, for example, operated to connect groups of people in one place (who had the finances) to those elsewhere in need (i.e. linkages based on 'sympathy' for others), increasingly people and places of the globe are being linked in empathy and solidarity. It has been seen that transnational social movements currently connect many individuals and varied networks and associations of different groups and organisations across countries and at the same time, through what are understood as inter-related and shared struggles and resistances in the spheres of environment, economy and social justice.

The chapters of the book have also presented evidence of *difference* and *disputed* understandings of the future challenges in sustainable development. It continues to be widely argued that the sustainable development agenda at the international level is *not* commonly shared, but rather, dominated by a more narrow and particularly 'northern' interest. There is a persistent concern that it continues to be issues of 'Global Environmental Change' (such as the implications of climate warming and biodiversity conservation on future generations) rather than international distributive issues of poverty that impact on current generations (and are much harder to tackle, are less amenable to 'mitigation' and technological solutions, for example) that dominate policy and action. The failure to find an international consensus on the shape of the Climate Convention after 2012 and the end of the current Kyoto Protocol was seen in Chapter 2 to encapsulate the different and disputed understandings of the climate challenge. Indeed, a key part of the dispute and contestation at the Copenhagen meeting in 2010 concerned how to take forward into the future the existing mechanism for recognising 'common but differentiated' responsibilities in action on climate change, that is, the legally binding targets for emissions reductions that to date referred only to Annex 1 countries. However, it also exposed how 'northern' and 'southern' understandings of the future challenges of action on climate change are also differentiated; the US adopting a very different stance on future targets to the European Union, for example, and as also seen between the BRIC economies and the Small Island states.

Evidence was also provided of research and practice that puts difference and inclusivity central to the future challenge of climate warming for sustainable development. The Intergovernmental Panel on Climate Change, for example, is working with the understanding that different 'emission scenarios' or 'pathways of development' (i.e. what happens in practice in economic, political and social terms) will be key in terms of influencing the scale and distribution of global and regional impacts of climate warming (Parry, 2004). Hence, it is understood that mitigation efforts will not prevent climate change from occurring and closer attention is needed to the challenges of adaptation. This includes understanding how development opportunities are being compromised through climate change, finding ways to integrate climate change adaptation throughout development strategies and policies and ensuring that future donor support (including as comes through the international carbon market) is focused on developing adaptation capacity in developing regions.

Disputed understandings of the notion of 'common futures' in the global challenges of sustainable development were also identified in Chapter 3 in the arena of trade. Whilst proponents of free trade (and neo-liberal development ideas generally) emphasise the global benefits that will flow from enhanced trade and economic activity, the campaigns of NGOs such as Oxfam are centred around how the international trade rules currently *don't* constitute a 'level playing field'. Their argument is that multilateral trade arrangements do not constitute a *common* starting point, nor do they present equal opportunities for development in the near future. The importance of a multilateral trade regime that is better aligned with the objectives of human development is also recognised by institutions including the UN Development Programme. In their report, 'Making Global Trade Work for People' the UNDP urged that the multilateral trade regimes needed to reflect the *differences* between developing countries and industrialised countries more effectively; to give policy space for the coexistence of diverse development strategies and to permit asymmetric rules that favour the weakest members where required. Developing countries needed to have greater flexibility in developing their agricultural policies to achieve food security and to protect the most vulnerable. In the arena of the environment, flexibility needs to be ensured so that developing countries are enabled to design appropriate solutions according to their own development and environment priorities and without fear of trade sanctions. However, as seen in the text, despite commitments of members of the World Trade Organisation in 2001 to a 'Development Round' of negotiations, putting an agenda into practice has been substantially stalled in recent years.

Perhaps most explicitly, the text has consistently revealed quite *different* environmental concerns that challenge less developed regions of the world in comparison to more industrialised countries. As Redclift (1992: 26) identified a decade ago:

> In urbanised, industrial societies, relatively few people's livelihoods
> are threatened by conservation measures. The 'quality of life'
> considerations which play such a large part in dictating the political
> priorities of developed countries surface precisely because of the
> success of industrial capitalism in delivering relatively high standards
> of living for the majority (but by no means all) of the population. In
> the South, on the other hand, struggles over the environment are
> usually about basic needs, cultural identity and strategies of survival.

Evidently, some things have changed since Redclift's observation. Through the text it has been seen that high rates of economic growth in countries within the Global South such as in China and India have also been very successful in moving large numbers of people out of extreme poverty and out of slum living and working conditions, for example. Many 'middle income' countries (as seen in the reporting of the Happy Planet Index above) have also been able to deliver improvements in well-being at much lower levels and rates of economic growth (and with lower resource demands). However, there are many more examples throughout the book of income levels remaining a key factor in shaping access to natural resources and to basic environmental improvements, in influencing the security of livelihoods and tenure in both rural and urban contexts and in explaining the opportunity that people have to live healthy lives that (as Sen (2000) suggests) they themselves have reason to value.

However, it has also been seen throughout the chapters, that more sustainable development processes are being built through addressing those diverse aspects of poverty that are not necessarily closely linked to income, but rather, are focused on the barriers to inclusion and development that come through a lack of access to information or a lack of voice and democracy. Furthermore, key to many of the interventions towards more sustainable processes and patterns of development as seen and to recognising diversity of interests and promoting inclusivity has been the importance of 'empowering the poor', 'finding community' and taking action closest to the people and environments involved. For example, the World Bank now requires widespread consultation of civil society groups within the process of preparation of Poverty Reduction Strategy Papers that shape further lending to and eligibility of countries for programmes of debt relief. PRSPs have been argued to reflect an acknowledgement on behalf of the World Bank that there can be no single prescription for macro-economic adjustment for recipient countries, but that policies need to be tailored to the specific conditions and needs of particular countries as defined by multiple stakeholders. PRSPs (in contrast to earlier Structural Adjustment Programmes) aim to provide national ownership of the process based on broad-based participation throughout and over the longer term of civil society in partnership with government, other country-stakeholders and donors. Bilateral and private aid organisations are also now working through multiple channels including NGOs rather than largely through governments as in the past, towards more flexible but also targeted financial support to countries and communities.

Participation and inclusivity of multiple voices, knowledges and capacities have also been identified as the cornerstones of the 'Farmer First' approach in research and development that has had widespread impact on more sustainable rural development outcomes. The 'decentralisation' of authority in terms of finances and decision-making from national departments to city authorities has also been seen, in Chapter 5, to be central to progress towards more sustainable urban developments particularly for the way in which it raises accountability to local populations. Substantial evidence has been presented through the text of various institutions in development 'finding' community to the benefit of local peoples and environments. The role of 'southern-based' NGOs in particular such as SEWA in Chapter 4 and SPARC in Chapter 5 have been recognised to bring substantial expertise and capacity in terms of working at the local level across a variety of sectors in sustainable development and to have provided momentum for further change.

However, all such steps towards 'working with community', whilst not depending necessarily on large transfers of financial resources, have been seen to rest on very significant shifts in political power. Working to transfer and/or create power and control at a local level and within communities depends on many interrelated actions across the hierarchy of levels of action towards sustainable development considered through the text. They require difficult political decisions and often profound (and unsettling) changes. This applies as much to consumers in the global North, professionals in development research and amongst government leaders in international and national negotiations, as it does to city planners, NGO staff, and to men and women in the communities of the developing world themselves.

Whilst community participation has become a pervasive alternative discourse in sustainable development, it does not offer an unproblematic panacea for sustainable development. As Cooke and Kothari (2001) have strongly articulated, the complexities of power and power relations can still be misunderstood and misused in interventions within this agenda; an enthusiasm for participatory methods in research can drive out others with advantages not delivered by those tools and there may well be decision-making processes that are not participatory that are equally valid in practice. Unjust or 'tyrannical' uses of power (to use their term) can characterise participatory development processes and practices as much as within those agendas they have sought to contest.

The achievements made towards sustainable development as detailed in this text have involved substantial reassessment on behalf of individuals, governments, civil society, private business and industry and international institutions of the constraints and opportunities of development and the environment, of the criteria for success in the use and protection of the resources of the world and of notions of human progress and well-being. As Reid (1995: 235) has stated:

Sustainable development confronts, not just society, but each of us at
the heart of his or her purpose. It invites us to give practical support to
the values of social equity, human worth and ecological health; it
questions our readiness to involve ourselves in the struggle for change.

There can be no 'blueprints' for sustainable development: sustainable
development actions depend on embracing complexity and working
to reconcile different interests in environment and development. It is
not possible to predict what the likely needs of future generations
will be, the nature of technological progress that will be made or the
precise outcomes of global warming. Flexible and creative solutions
are required as the nature of the 'problem' evolves and as policies,
programmes and projects proceed. However, these are not
justifications for inaction today. The obligation of the current
generation is both to use and to protect the resources of the world in
ways that meet human development opportunities more equitably
today, but which do not preclude options for such actions tomorrow.

Summary

The field of indictors of and for sustainable development is
expanding rapidly. The tools for assessing sustainable development
targets and outcomes are now substantial but there continue to be
challenges for capturing the process of change and the more
qualitative dimensions of sustainability.

The economic vulnerability of countries globally, as demonstrated
by the recent economic crisis, has prompted questioning of both
the measures for assessing development progress and the means for
achieving this. It is the focus of a number of international reviews
of indicators for sustainable development.

Ecological footprinting is a well-established method for exploring
human impacts on environmental resources. It forms the basis of
carbon footprinting and also measures that quantify the efficiency
with which ecological resources are converted into human
development outcomes.

Green GDP accounting seeks to adjust and/or augment conventional
GDP accounts to better measure sustainable economic welfare.

Values-based outcomes such as respect for ecological integrity and social
justice are 'harder to measure' but central to the mission and activities of
many organisations working towards sustainable development.

Discussion questions

- What are the benefits of moves to 'tie down and measure' sustainable development?

- Research in more depth how methods of ecological footprinting are being improved. Find out your own ecological footprint using the method established by the World Wildlife Fund; http://footprint.wwf.org.uk/

- What are the challenges of trying to measure human happiness?

- What are your own core values? How do they relate to the natural environment? How do they relate to future generations?

- Think about the communities in which you are involved. What evidence is there for commitment to common goals for the future?

Further reading

Bell, S. and Morse, S. (2008) *Sustainability Indicators: Measuring the Immeasurable?*, second edition, Earthscan, London. An established text that looks at the theory and practice behind the development of sustainability indicators. The second edition looks at some very advanced and complex practical approaches to sustainability analysis.

Jackson, T. (2009) *Prosperity without Growth: Economics for a Finite Planet*, Earthscan, London. This text is a very readable consideration of how to reconcile economic growth, human well-being and ecological health now and into the future. A good source to understand more about how human well-being has become decoupled from economic growth, the continued dependence of economic growth on energy and resources/materials and the challenges of changing consumption patterns.

New Economics Foundation (2009) *The (Un) Happy Planet Index: Why Good Lives Don't Have to Cost the Earth,* NEF, London. For further information on the methods underpinning the Index and explanation of the global patterns found.

Wilkinson, R. and Picket, K. (2009) *The Spirit Level – Why More Equal Societies Almost Always Do Better*, Penguin, London. This book draws together the mounting evidence globally of the benefits of equality and how it is inequality rather than how wealthy a society is that explains so many patterns and problems of development.

Websites

www.neweconomics.org   New Economics Foundation is an independent think tank that seeks innovative solutions and challenges mainstream ideas on environment, economy and society. It works across all sectors including government, academia and civil society, produces some very useful reviews of research and provides guidance for those institutions looking to promote more sustainable development.

www.globalreporting.org   Website for the Global Reporting Initiative and the research and policy advice that they provide for organisations looking to improve the quality and scope of their sustainability reporting.

http://sd.defra.gov.uk   For full details on how the UK government is developing policy and action and providing support towards more sustainable development.

www.earthcharterinaction.org   For details of the origin of the Earth Charter Initiative and how it is supporting people, organisations and communities to now take action on the initiative in arenas of education, business, the media, religion and in law-making for example.

References

Actionaid (2003) *GM Crops – Going Against the Grain*, available online at www.actionaid.org (accessed 6 February 2012).

Adams, W.M. (2001) *Green Development: Environment and Sustainability in the Third World*, second edition, Routledge, London.

Adams, W.M. (2009) *Green Development*, third edition, Routledge, Abingdon.

Adams, W.M. and Hulme, D. (2001) 'Conservation and community: changing narratives, policies and practices in African conservation', in Hulme, D. and Murphree, M. (eds) *African Wildlife and Livelihoods: The Promise and Performance of Community Conservation*, James Currey, Oxford, pp. 9–23.

Adger, W.N. and Jordan, A. (eds) (2009) *Governing Sustainability*, Cambridge University Press, Cambridge.

Adger, W.M., Huq, S., Brown, K., Conway, D. and Hulme, M. (2003) 'Adaptation to climate change in the developing world', *Progress in Development Studies*, 3,3, pp. 179–95.

Agnew, C. and Woodhouse, P. (2010) *Water Resources and Development*, Routledge, Abingdon.

African Development Bank (2003) *Poverty and Climate Change: Reducing the Vulnerability of the Poor Through Adaptation*.

Africa Progress Panel (2010) *From Agenda to Action: Turning Resources into Results for People*, Africa Progress Report, Geneva.

Agrawal, A. and Gibson, C.C. (1999) 'Enchantment and disenchantment: the role of community in natural resource conservation', *World Development*, 27,4, pp. 629–49.

Ainger, K. (2002) 'Earth Summit for sale', *New Internationalist*, 347, pp. 20–2.

Anand, S. and Sen, S. (2000) 'Human development and economic sustainability', *World Development*, 28,12, pp. 2029–49.

Arnell, N.W., Livermore, M.J.L., Kovats, S., Levy, P.E., Nicholls, R., Parry, M.L. and Gaffin, S.R. (2004) 'Climate and socio-economic scenarios for global-scale climate change impacts assessments: characterising the SRES storylines', *Global Environmental Change*, 14, pp. 3–20.

Atkins, P.J. and Bowler, I.R. (2001) *Food and Society: Economy, Culture, Geography*, Arnold, London.

Auty, R. (1993) *Sustaining Development in Mineral Economies: The Resource-Curse Thesis*, Routledge, London.

Bah, M., Cisse, S., Diyamett, B., Diallo, G., Lerise, F., Okali, D., Okpara, E., Olawoye, J. and Tacoli, C. (2006) 'Changing Rural–Urban linkages in Mali, Nigeria and Tanzania', in Tacoli, C. (ed.) *The Earthscan Reader in Rural–Urban Linkages*, Earthscan, London, pp. 56–67.

Bahra, P. (2009) 'David Attenborough to be patron of Optimum Population Trust', *The Times*, 14 April.

BAN (Basel Action Network) (2002) *Exporting Harm: The High-Tech Trashing of Asia*, Basel Action Network, Seattle.

Barbier, E.B. (1987) 'The concept of sustainable economic development', *Environmental Conservation*, 14,2, pp. 101–10.

Barr, S. (2008) *Environment and Society: Sustainability, Policy and the Citizen*, Ashgate, Aldershot.

Bartelmus, P. (1994) *Environment, Growth and Development: The Concepts and Strategies of Sustainable Development*, Routledge, London.

Bartone, C., Bernstein, J., Leitmann, J. and Eigen, J. (1994) *Towards Environmental Strategies for Cities*, Routledge, London.

Bass, S. and Dalal-Clayton, B. (2004) 'National sustainable development strategies', in Bigg, T. (ed.) *Survival for a Small Planet: The Sustainable Development Agenda*, Earthscan/IIED, London, pp. 101–20.

Bebbington, A. (2004) 'Movements and modernisations, markets and municipalities: indigenous federations in rural Ecuador', in Peet, R. and Watts, M. (eds) *Liberation Ecologies: Environment, Development, Social Movements*, second edition, Routledge, London, pp. 394–421.

Beder, S. (2002) *Global Spin: The Corporate Assault on Environmentalism*, revised edition, Green Books, Totnes, Devon.

Bell, S. and Morse, S. (2008) *Sustainability Indicators: Measuring the Immeasurable?*, second edition, Earthscan, London.

Berkhout, F., Leach, M. and Scoones, I. (eds) (2003) *Negotiating Environmental Change: New Perspectives from Social Science*, Edward Elgar, Cheltenham.

Bigg, T. (2004) 'The World Summit on Sustainable Development: was it worthwhile?', in Bigg, T. (ed.) *Survival for a Small Planet: The Sustainable Development Agenda*, Earthscan/IIED, London, pp. 3–22.

Biswas, A.K. (1993) 'Management of international waters', *International Journal of Water Resources Development*, 9,2, pp. 167–89.

Biswas, M.R. and Biswas, A.K. (1985) 'The global environment: past, present and future', *Resources Policy*, 11,1, pp. 25–42.

Black, M. (2008) 'We need to talk about toilets', *New Internationalist*, 414, pp. 4–7.

Blaikie, P. and Brookfield, H. (1987) *Land Degradation and Development*, Methuen, London.

Blewitt, J. (2008) *Understanding Sustainable Development*, Earthscan, London.

Bojo, J., Green, K., Kishore, S., Pilapitiya, S. and Chandra Reddy, R. (2004) *Environment in Poverty Reduction Strategies and Poverty Reduction Support Credits*, World Bank Environment Department, IBRD, Washington.

Bond, A.J. and Morrison-Saunders, A. (2011) 'Re-evaluating sustainability assessment: aligning the vision and the practice', *Environmental Impact Assessment Review*, 31, pp. 1–7.

Bown, W. (1994) 'Deaths linked to London smog', *New Scientist*, 25 June, p. 4.

Boyd, E., Hultman, N., Timmons Roberts, J., Corbera, E., Cole, J., Bozmoski, A., Ebeling, J., Tippman, R., Mann, P., Brown, K.A. and Liverman, D.M. (2009) 'Reforming the CDM for sustainable development: lessons learned and policy futures', *Environmental Science and Policy*, 12,7, pp. 820–31.

Braidotti, R., Charkiewicz, E., Hausler, S. and Wieringa, S. (1994) *Women, the Environment and Sustainable Development: Towards a Theoretical Synthesis*, Zed Books, London.

Bryceson, D.F. (2002) 'The scramble in Africa: reorienting rural livelihoods', *World Development*, 30,5, pp. 725–39.

Buckles, D. and Rusnak, G. (1999) 'Conflict and collaboration in natural resource management', in Buckles, D. (ed.) *Cultivating Peace: Conflict and Collaboration in Natural Resource Management*, International Development Research Centre, Ottawa, Canada, pp. 1–10.

Buckley, C. (2010) 'China's farmers mount movement against land grabs', Reuters, 3 August.

Building Bridges Collective (2010) *Space for Movement? Reflections from Bolivia on Climate Justice, Social Movements and the State*, Footprint Workers Co-op, Leeds.

Burningham, D. and Davies, J. (1999) *Green Economics*, second edition, Heinemann, Oxford.

Cairncross, F. (1995) *Green Inc.: A Guide to Business and the Environment*, Earthscan, London.

Carlsen, L. (2004) 'The people of the corn', *New Internationalist*, 374, pp. 12–13.

Carter, N. (2007) *The Politics of the Environment: Ideas, Activism, Policy*, second edition, Cambridge University Press, Cambridge.

Castree, N. (2003) 'Uneven development, globalisation and environmental change', in Morris, D., Freeland, J., Hinchliffe, S. and Smith, S. (eds) *Changing Environments*, Oxford University Press, Oxford, pp. 275–312.

Center for Human Rights and Global Justice (2011) *Every Thirty Minutes: Farmer Suicides, Human Rights and the Agrarian Crisis in India*, NYU School of Law, New York.

Chambers, N., Simmons, C. and Wackernagel, M. (2000) *Sharing Nature's Interest: Ecological Footprints as an Indicator of Sustainable Development*, Earthscan, London.

Chambers, R. (1983) *Rural Development: Putting the Last First*, Longman, London.

Chambers, R. (1993) *Challenging the Professions: Frontiers for Rural Development*, Intermediate Technology Publications, London.

Chambers, R. (2008) *Revolutions in Development Inquiry*, Earthscan, London.

Chambers, R. and Conway, G. (1992) *Sustainable Rural Livelihoods: Practical Concepts for the Twenty-first Century,* IDS Discussion Paper 296, Brighton.

Chambers, R., Pacey, A. and Thrupp, L.A. (1989) *Farmer First: Farmer Innovation and Agricultural Research*, Intermediate Technology Publications, London.

Christian Aid (2004) *Fuelling Poverty*, Christian Aid, London.

Christian Aid (2007) *Human Tide: The Real Migration Crisis*, Christian Aid, London.

Clarke, R. and King, J. (2004) *The Atlas of Water: Mapping the World's Most Critical Resource*, Earthscan, London.

Clayton, K. (1995) 'The threat of global warming', in O'Riordan, T. (ed.) *Environmental Science for Environmental Management*, Longman, London, pp. 110–31.

Conroy, C. and Litvinoff, C. (1988) *The Greening of Aid: Sustainable Livelihoods in Practice*, Earthscan, London.

Conway, G.R. (1987) 'The properties of agroecosystems', *Agricultural Systems*, 24, pp. 95–117.

Conway, G.R. (1997) *The Doubly Green Revolution: Food for All in the 21st Century*, Penguin, London.

Cooke, B. and Kothari, U. (eds) (2001) *Participation: The New Tyranny?* Zed Books, London.

Corbridge, S. (1987) 'Development and underdevelopment', *Geography Review*, September, pp. 20–2.

Corbridge, S. (1999) 'Development, post-development and the global political economy', in Cloke, P., Crang, P. and Goodwin, M. (eds) *Introducing Human Geographies*, Arnold, London, pp. 67–75.

Cotula, L., Vermeulen, S., Leonard, R. and Keeley, J. (2009) *Land Grab or Development Opportunity: Agricultural Investment and International Land Deals in Africa,* IIED/FAO/IFAD, London/Rome.

Craig, G. and Mayo, M. (eds) (1995) *Community Empowerment: A Reader in Participation and Development*, Zed Books, London.

Curto, S. (2007) 'This changing aid landscape', *Finance and Development*, December, pp. 20–1.

Dahl, A.L. (2012) 'Achievements and gaps in indicators for sustainability', *Ecological Indicators*, 17, pp. 14–19.

Davis, B., Winters, P. and Carletto, G. (2010) 'A cross-country comparison of rural income generating activities', *World Development*, 38,1, pp. 48–63.

DEFRA (2005) *Securing the Future: Delivering the UK Government Sustainable Development Strategy*, Department for Environment, Food and Rural Affairs, London.

DEFRA (2008) *Review of Work on the Environmental Sustainability of International Biofuels Production and Use*, Department for Environment, Food and Rural Affairs, London.

DEFRA (2009) *Sustainable Development Indicators in your Pocket*, Department for Environment, Food and Rural Affairs, London.

Desai, V. and Potter, R.B. (eds) (2008) *The Companion to Development Studies*, second edition, Hodder Education, London.

DETR (1999) *Quality of Life Counts. Indicators for a Strategy for Sustainable Development in the UK: A Baseline Survey*, Department of the Environment, Transport and the Regions, London.

Deutscher, E. and Fyson, S. (2008) 'Improving the effectiveness of aid', *Finance and Development*, September, pp. 15–19.

Devereux, S. (1993) 'Goats before ploughs: dilemmas of household response sequencing during food shortages', *IDS Bulletin*, 24,4, pp. 52–9.

Devereux, S. (1999) *Theories of Famine*, Harvester Wheatsheaf, New York.

Dicken, P. (2011) *Global Shift: Mapping the Changing Contours of the World Economy*, sixth edition, Sage, London.

Dobson, A. (2009) 'Citizens, citizenship and governance for sustainability', in Adger, W.N. and Jordan, A. (eds) *Governing Sustainability*, Cambridge University Press, Cambridge, pp. 125–41.

Driscoll, R. and Evans, A. (2005) 'Second-generation poverty reduction strategies: new opportunities and emerging issues', *Development Policy Review*, 23,1, pp. 5–25.

Economic Commission for Europe (1998) *Convention on Access to Information, Public Participation in Decision-making and Access to Justice in Environmental Matters*, ECE, Aarhus, Denmark.

Edge, G. and Tovey, K. (1995) 'Energy: hard choices ahead', in O'Riordan, T. (ed.) *Environmental Science for Environmental Management*, Longman, London, pp. 317–34.

Edwards, M. and Gaventa, J. (eds) (2001) *Global Citizen Action*, Earthscan, London.

Ehrlich, P. (1968 [reprinted 1995]) *The Population Bomb*, Buccaneer Books, New York.

Ekins, P. (2003) 'After Seattle: what next for trade and the environment?', in Berkhout, F., Leach, M. and Scoones, I. (eds) *Negotiating Environmental Change: New Perspectives from Social Science*, Edward Elgar, Cheltenham, pp. 159–92.

Elliott, J.A. (2002) 'Towards sustainable rural resource management in sub-Saharan Africa', *Geography*, 87,3, pp. 197–204.

Ellis, F. (2000) *Rural Livelihoods and Diversity in Developing Countries*, Oxford University Press, Oxford.

Energy Information Administration (2006) *International Energy Annual 2006*, available online at http://www.eia.gov/cfapps/ipdbproject/IEDIndex3.cfm (accessed 7 March 2012).

Energy Information Administration (2009) *International Energy Statistics 2009*, available online at http://www.eia.gov/cfapps/ipdbproject/IEDIndex3.cfm (accessed 7 March 2012).

Escobar, A. (1995) *Encountering Development: The Making and Unmaking of the Third World*, Princeton University Press, Princeton.

ESDinds Project (2010) *We Value: Understanding and Evaluating the Intangible Aspects of your Work*, University of Brighton, Brighton.

Evans, J.P. (2012) *Environmental Governance*, Routledge, London.

Evans, R. (2010) 'Oil company Trafigura fined Euros1 million for sending toxic waste to Africa', *Guardian*, 24 July.

FAO (Food and Agriculture Organisation) (2007) Profitability and sustainability of urban and peri-urban agriculture. Agricultural management, marketing and finance occasional paper 19. FAO, Rome.

FAO (2008) *The State of Food and Agriculture 2008, Biofuels: Prospects, Risks and Opportunities*, FAO, Rome.

FAO (2010) The State of Food Insecurity in the World: Addressing Insecurity in Protracted Crises, FAO, Rome.

Finger, M. (1994) 'Environmental NGOs in the UNCED process', in Princen, T. and Finger, M. (eds) *Environmental NGOs in World Politics*, Routledge, London, pp. 186–213.

Flavin, C. (2008) 'Building a low-carbon economy', in *State of the World 2008: Innovations for a Sustainable Economy*, Earthscan, London, pp. 75–90.

Flavin, C. and Gardner, G. (2006) 'China, India and the New World Order', in *State of the World 2006: The Challenge of Global Sustainability*, Earthscan, London, pp. 3–23.

FOE (Friends of the Earth) (1999) *The IMF: Selling the Environment Short*, Friends of the Earth, Washington, also available online at www.foe.org/international (accessed 6 February 2012).

FOE (2010) *Redd: The Realities in Black and White*, FOEI secretariat, Amsterdam, also available online at http://foei.org/redd-realities (accessed 6 February 2012).

Ford, L.H. (1999) 'Social movements and the globalisation of environmental governance', *IDS Bulletin*, 30,3, pp. 68–74.

Forum for the Future (2010) *The Future Climate for Development: Scenarios for Low-income Countries in a Climate-changing World*, Forum for the Future/DFID, London.

Frank, A.G. (1967) *Capitalism and Underdevelopment in Latin America*, Monthly Review Press, New York.

FTLOI (Fair Trade Labelling Organizations International) (2010) *Growing Stronger Together: Annual Report 2009–10*, Fair Trade Labelling Organizations International, Bonn Germany.

George, S. (1992) *The Debt Boomerang*, Pluto Press, London.

Gibson, R.B. (2005) *Sustainability Assessment*, Earthscan, London.

Giddens, A. (2009) *The Politics of Climate Change*, Polity Press, Cambridge.

Glennie, J. (2008) *The Trouble with Aid: Why Less Could Mean More for Africa*, Zed Books, London.

Glennie, J. (2009) 'Openness, democracy and humility: the principles of a new era in development', *Development*, 52,3, pp. 363–8.

The Global Fund (no date) *Debt2Health: Innovative Financing of the Global Fund*, available online at http://www.theglobalfund.org/en/innovative financing/documentation/ (accessed 21 September 2011).

Goldsmith, E., Allen, R., Allaby, M., Davoll, J. and Lawrence, S. (eds) (1972) *Blueprint for Survival*, Penguin, Harmondsworth.

Goodman, J. and Wright, M. (2011) '2011: five trends to watch', *Green Futures*, 79, pp. 16–17.

Gould, W.T.S. (2009) *Population and Development*, Routledge, Abingdon.

Greenpeace (2010) *Koch industries secretly funding the climate denial machine*, published by Greenpeace USA, Washington, also available online at http://www.greenpeace.org/usa/en/media-center/reports/executive-summary-koch-indus/ (accessed 7 March 2012).

Griffiths, T. (2007) *Seeing 'Red'? 'Avoided Deforestation' and the Rights of Indigenous Peoples and Local Communities*, Forest Peoples Programme, Moreton-in-Marsh, UK.

Guijt, I. and Shah, M.K. (1998) *The Myth of the Community: Gender Issues in Participatory Development*, Intermediate Technology Publications, London.

Gupta, A. (1998) *Ecology and Development in the Third World*, second edition, Routledge, London.

Gwynne, R.N. (2008) 'Free trade and fair trade' in Desai, V. and Potter, R.B. (eds) *The Companion to Development Studies*, second edition, Arnold, London, pp. 201–6.

Hamilton, K., Sjardin, M. Peters-Stanley, M. and Marcellos, T. (2010) *State of the Voluntary Carbon Market, 2010,* Ecosystem Marketplace, Washington and Bloomberg New Energy Finance, New York.

Hardoy, J.E. and Satterthwaite, D. (1989) *Squatter Citizen: Life in the Urban Third World*, Earthscan, London.

Hardoy, J.E., Mitlin, D. and Satterthwaite, D. (2001) *Environmental Problems in an Urbanising World*, Earthscan, London.

Hettne, B. (2002) 'Current trends and future options in development studies', in Desai, V. and Potter, R.B. (eds) *The Companion to Development Studies*, Arnold, London, pp. 7–11.

Hewitt, T. (2000) 'Half a century of development', in Allen, T. and Thomas, A. (eds) *Poverty and Development into the Twenty-first Century*, Oxford University Press, Oxford, pp. 289–308.

Hildyard, N. (1994) 'The big brother bank', *Geographical Magazine*, June, pp. 26–8.

Hill, A.G. (1991) 'African demographic regimes: past and present', paper presented at the Conference of the Royal African Society, Cambridge, April.

Hodder, R. (2000) *Development Geography*, Routledge, London.

Houghton, J. (2009) *Global Warming: The Complete Briefing*, fourth edition, Cambridge University Press, Cambridge.

Huggler, J. (2004) 'The price of dignity for Fatima?', *Independent*, 9 December.

IAASTD (International Assessment of Agricultural Knowledge, Science and Technology for Development) (2009) *Agriculture at a Cross-roads: Synthesis Report*, Island Press, Washington.

IEA (International Energy Agency) (2009) *World Energy Outlook*, IEA, Paris.

IEA (2011) *Key World Energy Statistics*, IEA, Paris.

IIED (International Institute for Environment and Development) (2007) *Up in Smoke? Asia and the Pacific: The Threat from Climate Change to Human Development and the Environment*, available online at www.iied.org/pubs/ (accessed 23 September 2008).

IIED (2009) 'Community-based adaptation to climate change', *Participatory Learning and Action 60*, IIED, London.

IISD (International Institute for Sustainable Development) (no date) *BellagioSTAMP*, Measurement and Assessment Program, Manitoba, available online at www.iisd.org/measure/ (accessed 8 December 2010).

IMF (International Monetary Fund) (2010) Poverty reduction strategy papers, *IMF Factsheet*, available online at www.imf.org/external/np/exr/facts/prsp. htm (accessed 29 September 2011).

IPCC (Intergovernmental Panel on Climatic Change) (1995) *Climate Change 1995: Second assessment*, IPCC, Geneva.

IPCC (2001) *Climate Change 2001: Impacts, Adaptation and Vulnerability*, IPCC, Geneva.

IPCC (2007) *Climate Change 2007: The Physical Science Basis: Technical Summary*, IPCC, Geneva.

IUCN (International Union for Conservation of Nature) (1980) *World Conservation Strategy*, International Union of Conservation of Nature and Natural Resources, UNEP, WWF, Geneva.

Jackson, T. (2008) 'The challenge of sustainable lifestyles', in Worldwatch Institute, *State of the World, 2008: Ideas and Opportunities for Sustainable Economies*, Earthscan, London, pp. 45–60.

Jackson, T. (2009) *Prosperity without Growth: Economics for a Finite Planet*, Earthscan, London.

Jacobs, M. (1991) *The Green Economy*, Pluto Press, London.

Jacobs, S. (2010) *Gender and Agrarian Reforms*, Routledge, Abingdon.

James, C. (2010) 'Global status of commercialised biotech/GM crops 2010', *ISSA Brief no 42*, International Service for the Acquisition of Agri-biotech Applications, Ithaca, New York.

Johansson, P. (2010) 'Debt relief, investment and growth', *World Development*, 38,9, pp. 1204–16.

Johnston-Hernandez, B. (1993) 'Dirty growth', *New Internationalist*, 246, pp. 10–11.

Jones, G.A. and Corbridge, S. (2010) 'The continuing debate about urban bias: the thesis, its critics, its influence and its implications for poverty-reduction strategies', *Progress in Development Studies*, 10,1, pp. 1–18.

Jordan, A. and Brown, K. (2007) 'The international dimensions of sustainable development: Rio reconsidered', in Auty, R.M. and Brown, K. (eds) *Approaches to Sustainable Development*, Pinter, London, pp. 270–95.

Joshi, D., Fawcett, B. and Mannan, F. (2011) 'Health, hygiene and appropriate sanitation: experiences and perceptions of the urban poor', *Environment and Urbanization*, 23,1, pp. 91–111.

Knox, P.L. and Marston, S.A. (1998) *Places and Regions in Global Context*, second edition, Prentice-Hall, New Jersey.

Knox, P.L. and Martson, S.A. (2009) *Human Geography: Places and Regions in a Global Context*, fifth edition, Pearson Education International, Harlow.

Korten, D.C. (1990) *Getting to the Twenty-First Century: Voluntary Action and the Global Agenda*, Kumarian Press, New York.

Korten, D.C. (2001) 'The responsibility of business to the whole', in Starkey, R. and Walford, R. (eds) *The Earthscan Reader in Business and Sustainable Development*, Earthscan, London, pp. 230–41.

Leach, M. and Mearns, R. (1991) *Poverty and Environment in Developing Countries: An Overview Study*, final report to ESRC and ODA, Institute of Development Studies, University of Sussex, Brighton.

Lee, K., Holland, A. and McNeill, D. (eds) (2000) *Global Sustainable Development in the Twenty-First Century*, Edinburgh University Press, Edinburgh.

Lemos, M.C. and Agrawal, A. (2006) 'Environmental Governance', *Annual Review of Environmental Resources*, 31, pp. 297–325.

Leonard, H.J. (1989) *Environment and the Poor: Development Strategies for a Common Agenda*, Transaction Books, Oxford.

Levett, R. (1998) 'Footprinting: a great step forward, but tread carefully – a response to Mathis Wackernagel', *Local Environment*, 3,2, pp. 67–74.

Lipton, M. (1977) *Why Poor People Stay Poor: Urban Bias in World Development*, Temple Smith, London.

Lynch, K. (2005) *Rural–Urban Interactions in the Developing World*, Routledge, London.

Madeley, J. (2002) *Food for All: The Need for a New Agriculture*, Zed Books, London.

Malena, C. (2000) 'Beneficiaries, mercenaries, missionaries and revolutionaries: unpacking NGO involvement in World Bank financed projects', *IDS Bulletin*, 31,3, pp. 19–34.

Marray, M. (1991) 'Natural forgiveness', *Geographical Magazine*, 63,12, pp. 18–22.

Martinez-Alier, J. (2002) *The Environmentalism of the Poor: A Study of Ecological Conflicts and Valuation*, Edward Elgar, Cheltenham.

Mastny, L. and Cincotta, R.P. (2005) 'Examining the connections between population and security, in *State of the World 2005: Global Security*, World Resources institute, Earthscan, London, pp. 23–39.

Mather, A.S. and Chapman, K. (1995) *Environmental Resources*, Longman, London.

Mawhinney, M. (2002) *Sustainable Development: Understanding the Green Debates*, Blackwell Science, Oxford.

McCormick, J. (1995) *The Global Environment Movement*, second edition, Wiley, London.

McGranahan, G. and Murray, F. (eds) (2003) *Air Pollution and Health in Rapidly Developing Countries*, Earthscan, London.

McGranahan, G., and Satterthwaite, D. (2002) 'The environmental dimensions of sustainable development for cities', *Geography*, 87,3, pp. 213–26.

McGranahan, G., Satterthwaite, D. and Tacoli, C. (2004) *Rural–urban Change, Boundary Problems and Environmental Burdens*, IIED Working paper series on Rural–Urban interactions and livelihood strategies, Working Paper 10, IIED, London.

McGranahan, G., Jacobi, P., Songsore, J., Surjadi, C. and Kjellen, M. (2001) *The Citizens at Risk: From Urban Sanitation to Sustainable Cities*, Earthscan, London.

McNeill, D. (2000) 'The concept of sustainable development', in Lee, K., Holland, A. and McNeill, D. (eds) *Global Sustainable Development in the Twenty-first Century*, Edinburgh University Press, Edinburgh, pp. 10–29.

MEA (Millennium Ecosystem Assessment) (2005) *Ecosystems and Human Well-Being: Synthesis*, Island Press, Washington.

Meadows, D.H., Meadows, D.L., Randers, J. and Behrens III, W.W. (1972) *The Limits to Growth*, Universe Books, New York.

Mee, L.D., Dublin, H.T. and Eberhard, A.A. (2008) 'Evaluating the global environment facility: a goodwill gesture or a serious attempt to deliver global benefits?', *Global Environmental Change*, 18, pp. 800–10.

Middleton, N., O'Keefe, P. and Moyo, S. (1993) *Tears of the Crocodile: From Rio to Reality in the Developing World*, Pluto Press, London.

Milanovic, B. (2005) *Worlds Apart: Measuring International and Global Inequality*, Princeton University Press, Oxford.

Mitchell, R. (2003) 'International environmental agreements: a survey of their features, formation and effects', *Annual Review of Energy and the Environment*, 28, pp. 429–61.

Mitlin, D. and Satterthwaite, D. (2004) *Empowering Squatter Citizens: Local Government, Civil Society and Urban Poverty Reduction*, Earthscan, London.

Mohan, G. (2008) 'Participatory development in practice', in Desai, V. and Potter, R.B. (eds) *The Companion to Development Studies*, second edition, Arnold, London, pp. 45–9.

Mohan, G., Brown, E., Milward, B. and Zack-Williams, A.B. (2000) *Structural Adjustment: Theory, Practice and Impacts*, Routledge, London.

Momsen, J. (2010) *Gender and Development*, second edition, Routledge, Abingdon.

Morse, S. and Mannion, A.M. (2009) 'Can genetically modified cotton contribute to sustainable development in Africa?', *Progress in Development Studies*, 9,3, pp. 225–47.

Mortimore, M. (1998) *Roots in the African Dust: Sustaining the Drylands*, Cambridge University Press, Cambridge.

Moyo, D. (2009) *Dead Aid: Why Aid is Not Working and How there is a Better Way for Africa*, Farrar, Straus and Giroux, New York.

Mullen, J. (2008) 'Rural poverty', in Desai, V. and Potter, R.B. (eds.) *The Companion to Development Studies*, second edition, Hodder Education, London, pp. 143–6.

Myers, N. and Myers, D. (1982) 'Increasing awareness of the supranational nature of emerging environmental issues', *Ambio*, 11,4, pp. 195–201.

Naim, M. (2007) 'Rogue aid', *Foreign Policy*, 159, pp. 95–6.

Narayan, D. with Patel, R., Schafft, K., Rademacher, A. and Koch-Schulte, S. (2000) *Voices of the Poor: Can Anyone Hear Us?* Oxford University Press, Oxford.

Nayyar, D. (2008) 'Learning to Unlearn from Development', *Oxford Development Studies*, 36,3, pp. 259–80.

Neefjes, K. (2000) *Environments and Livelihoods: Strategies for Sustainability*, Oxfam, Oxford.

NEF (New Economics Foundation) (2006) *The UK Interdependence Report: How the World Sustains the Nation's Lifestyles and the Price it Pays*, NEF, London.

NEF (2008) *Debt Relief as if Justice Mattered*, NEF, London.

NEF (2009a) *The (Un) Happy Planet Index: Why Good Lives Don't Have to Cost the Earth*, NEF, London.

NEF (2009b) *Seven Principles for Measuring what Matters: A Guide to Effective Public Policy Making*, NEF, London.

Nelson, P.J. (2008) 'The World Bank and NGOs', in Desai, V. and Potter, R.B. (eds) *The Companion to Development Studies*, second edition, Hodder Education, London, pp. 550–4.

New Internationalist (2002) 'Inside business: how corporations make the rules', 347.

New Internationalist (2006) 'CO2NNED: Carbon offsets stripped bare', 391.

Nikiforuk, A. (2010) 'Canada's curse', *New Internationalist*, 431, pp. 8–10.

Obsibanjo, O. and Nnorom, I.C. (2007) 'The challenge of electronic waste management in developing countries', *Waste Management and Research*, 25, pp. 489–501.

OECD (2010a) *Development Co-operation Report 2010*, Development Assistance Committee, OECD Publishing, doi: 10.1787/dcr-2010-en.

OECD (2010b) *Aid in Support of Environment*, OECD-DAC Secretariat Creditor reporting System Database, available online at http://www.oecd.org/dataoecd/48/22/45078749.pdf (accessed 6 March 2012).

OECD (2010c) *Tracking Aid in Support of Climate Change Mitigation and Adaptation in Developing Countries*, available online at www.oecd,org/dac/stats/rioconventions (accessed 11 February 2011).

OECD (2010d) *OECD Factbook: Economic, Environmental and Social Statistics*, OECD Publishing, doi: 10.1787/fb10-data-en.

ONS (Office for National Statistics) (2007) 'Variations persist in life expectancy by social class', News Release 24 October, ONS, London.

O'Riordan, T. (1995) *Environmental Science for Environmental Management*, Longman, London.

O'Riordan, T. (ed.) (2000) *Environmental Science for Environmental Management*, second edition, Pearson Education, Harlow.

O'Riordan, T. (2009) 'Reflections on the pathways to sustainability', in Adger, W.N. and Jordan, A. (eds) *Governing Sustainability*, Cambridge University Press, Cambridge, pp 307–28.

Osibanjo, O. and Nnorom, I.C. (2007) 'The challenge of electronic waste (e-waste) management in developing countries', *Waste Management and Research*, 25, pp. 489–501.

Oxfam (2002) *Rigged Rules and Double Standards: Trade, Globalisation and the Fight against Poverty*, available online at www.maketradefair.com (accessed 21 July 2005).

Parry, M. (2004) 'Viewpoint: global impacts of climate change under the SRES scenarios', *Global Environmental Change*, 14, p. 1.

Peake, S. and Smith, J. (2009) *Climate Change: From Science to Sustainability*, second edition, Oxford University Press, Oxford.

Pearce, D., Markandya, A. and Barbier, E. (1989) *Blueprint for a Green Economy*, Earthscan, London.

Pearson, R. (2001) 'Rethinking gender matters in development', in Allen, T. and Thomas, A. (eds) *Poverty and Development into the 21st Century*, Oxford University Press, Oxford, pp. 383–402.

Peet, R. and Watts, M. (2004) *Liberation Ecologies: Environment, Development, Social Movements*, second edition, Routledge, London.

Pepper, D. (1984) *The Roots of Modern Environmentalism*, Croom Helm, London.

Pezzullo, P.C. and Sandler, R. (2007) 'Revisiting the environmental justice challenge to environmentalism', in Sandler, R. and Pezzullo, P.C. (eds) *Environmental Justice and Environmentalism*, The MIT Press, Cambridge, MA pp. 1–24.

Porritt, J. (2011) 'Is chronic political procrastination hampering the most progressive business leaders?', *Green Futures*, 21 February.

Potter, R.B. and Lloyd-Evans, S. (1998) *The City in the Developing World*, Longman, Harlow.

Potter, R.B., Binns, J.A., Elliott, J.A. and Smith, D. (2008) *Geographies of Development*, third edition, Pearson Education Limited, Harlow.

Pretty, J. (1995) *Regenerating Agriculture: Policies and Practices for Sustainability and Self-Reliance*, Earthscan, London.

Pretty, J. (2008) 'Agricultural sustainability', in Desai, V. and Potter, R.B. (eds) *The Companion to Development Studies*, second edition, Hodder Education, London, pp. 165–9.

Pretty, J. and Ward, H. (2001) 'Social capital and the environment', *World Development*, 29,2, pp. 209–27.

Rakodi, C. (2002) 'Economic development, urbanisation and poverty', in Rakodi, C. and Lloyd-Jones, T. (eds) *Urban Livelihoods: A People-Centred Approach to Reducing Poverty*, Earthscan, London, pp. 23–34.

Rapley, J. (2008) 'End of development or age of development?', *Progress in Development Studies*, 8,2, 177–82.

Redclift, M. (1987) *Sustainable Development: Exploring the Contradictions*, Routledge, London.

Redclift, M. (1992) 'Sustainable development and popular participation: a framework for analysis', in Ghai, D. and Vivian, J.M. (eds) *Grassroots Environmental Action: People's Participation in Sustainable Development*, Routledge, London, pp. 23–49.

Redclift, M. (1997) 'Sustainable development: needs, values, rights', in Owen, L. and Unwin, T. (eds) *Environmental Management: Readings and Case Studies*, Blackwell, London, pp. 438–50.

Redclift, M. (2005) 'Sustainable development (1987–2005): an oxymoron comes of age', *Sustainable Development*, 13, pp. 212–27.

Redefining Progress (2006) *The Genuine Progress Indicator 2006*, Redefining Progress, Oakland California, available online at www.redefiningprogress.org (accessed 6 February 2012).

Reed, D. (ed.) (1996) *Structural Adjustment, the Environment and Sustainable Development*, Earthscan, London.

Reid, D. (1995) *Sustainable Development: An Introductory Guide*, Earthscan, London.

Reiterer, M. (2009) 'The Doha development agenda of the WTO: possible institutional implications', *Progress in Development Studies*, 9,4, pp. 359–75.

Rich, B. (1994) *Mortgaging the Earth: The World Bank, Environmental Impoverishment and the Crisis of Development*, Earthscan, London.

Riddell, R.C. (2007) *Does Foreign Aid Really Work?*, Oxford University Press, Oxford.

Rigg, J. (1997) *SouthEast Asia*, Routledge, London.

Rigg, J. (2007) *An Everyday Geography of the Global South*, Routledge, Abingdon.

Robinson, G.M. (2004) *Geographies of Agriculture: Globalisation, Restructuring and Sustainability*, Pearson Education, Harlow.

Robinson, N. (ed.) (1993) *Agenda 21: Earth's Action Plan*, IUCN Environmental Policy and Law Paper 27, Oceana Publications, New York.

Rostow, W. (1960) *The Stages of Economic Growth: A Non-Communist Manifesto*, Cambridge University Press, Cambridge.

Sachs, J. (2008) *Common Wealth: Economics for a Crowded Planet*, Penguin, London.

Satterthwaite, D. (2008a) 'Urbanisation in developing countries', in Desai, V. and Potter, R.B. (eds) *The Companion to Development Studies*, second edition, Arnold, London, pp. 237–43.

Satterthwaite, D. (2008b) 'Urbanisation and environment in the third world', in Desai, V. and Potter, R.B. (eds) *The Companion to Development Studies*, second edition, Arnold, London, pp. 262–8.

Satterthwaite, D. and Tacoli, C. (2002) 'Seeking an understanding of poverty that recognises rural–urban differences and rural–urban linkages', in Rakodi, C. and Lloyd-Jones, T. (eds) *Urban Livelihoods: A People-Centred Approach to Reducing Poverty*, Earthscan, London, pp. 52–70.

Satterthwaite, D., Hart, R., Levy, C., Mitlin, D., Ross, D., Smit, J. and Stephens, C. (1996) *The Environment for Children: Understanding and Acting on the Environmental Hazards that Threaten Children and Their Parents*, Earthscan, London.

Sauer, S. (2009) 'Market led "agrarian reform" in Brazil: a dream has become a debt burden', *Progress in Development Studies*, 9,2, pp. 127–40.

Schipper, E.L.F (2007) *Climate Change Adaptation and Development: Exploring the Linkages*, Tyndal Centre for Climate Change Research Working Paper 107.

Schurmann, F.J. (2008) 'The impasse in development studies', in Desai, V. and Potter, R.B. (eds) *The Companion to Development Studies*, second edition, Arnold, London, pp. 12–16.

Scoones, I. and Thompson, J. (eds) (1994) *Beyond Farmer First: Rural People's Knowledge, Agricultural Research and Extension Practice*, IT Publications, London.

Scoones, I. and Thompson, J. (eds) (2009) *Farmer First Revisited: Innovation for Agricultural Research and Development*, Practical Action Publishing, Rugby.

SELCO (no date) 'Access to sustainable energy services via innovative financing', SELCO in collaboration with renewable energy and Energy Efficiency partnerships Vienna, Austria.

Sen, A. (1999) *Development as Freedom*, Oxford University Press, Oxford.

Sen, A.K. (1981) *Poverty and Famine: An Essay on Entitlement and Deprivation*, Clarendon, Oxford.

Seyfang, G. (2003) 'Environmental mega-conferences: from Stockholm to Johannesburg and beyond', *Global Environmental Change*, 13, pp. 223–8.

Shiva, V. (1989) *Staying Alive*, Zed Books, London.

Shiva, V. (2000) *Stolen Harvest: The Hijacking of the Global Food Supply*, Zed Books, London.

Silvey, R. and Elmhirst, R.E (2003) 'Engendering social capital: women workers and rural–urban networks in Indonesia's crisis', *World Development* 31,5, 865–79.

Simon, D. (2008) 'Neo-liberalism, structural adjustment and poverty reduction strategies', in Desai, V. and Potter, R.B. (eds) *The Companion to Development Studies*, second edition, Arnold, London, pp. 86–92.

Simpson, A. and Tuxworth, B. (2010) 'Creativity begins when you cut a zero from your budget', in Forum for the Future Sol e Sombra: Brazil's Search for a Sustainable Future, *Green Futures Special Publication*, pp. 22–3, Forum for the Future, London.

Starke, L. (1990) *Signs of Hope: Working Towards Our Common Future*, Oxford University Press, Oxford.

Starkey, R. and Walford, R. (eds) (2001) *The Earthscan Reader in Business and Sustainable Development*, introduction by the editors, 'Defining sustainable development', Earthscan, London.

Steffen, W., Sanderson, A., Tyson, P.D., Jager, J., Matson, P.A., Moore, B. III, Oldfield, F., Richardson, K., Schellenhuber, H.J., Turner, B.L. II, Wasson, R.J. (2004) *Global Change and the Earth System: A Planet Under Pressure*, Springer, New York.

Stern, N. (2007) *The Economics of Climate Change: The Stern Review*, Cambridge University Press, Cambridge.

Stern, N. (2009) *A Blueprint for a Safer Planet*, Bodley Head, London.

Stiglitz, J.E. (2009) 'The imperative for improved global economic coordination', *Development Outreach*, December, pp. 39–42.

Stiglitz, J.E., Sen, A. and Fitoussi, J-P. (2009) *Report by the Commission on the Measurement of Economic Performance and Social Progress*, available online at www.stiglitz-sen-fitoussi.fr (accessed 6 February 2012).

Stohr, W.B. and Taylor, D.R.F. (1981) *Development from Above or Below? The Dialectics of Regional Planning Developing Countries*, Wiley, Chichester.

Tacoli, C. (2003) 'The links between urban and rural development', *Environment and Urbanization*, 15,1, pp. 3–12.

Tacoli, C. (ed.) (2006) *The Earthscan Reader in Rural–Urban Linkages*, Earthscan, London.

Tacoli, C. (2009) 'Crisis or adaptation? Migration and climate change in a context of high mobility', *Environment and Urbanization*, 21,2, pp. 513–25.

Talberth, J. (2008) 'A new bottom line for progress', in Worldwatch Institute, *State of the World, 2008: Ideas and Opportunities for Sustainable Economies*, Earthscan, London, pp. 18–31.

Taylor, A. (2003) 'Trading with the environment', in Bingham, N., Blowers, A. and Belshaw, C. (eds) *Contested Environments*, Wiley, Chichester, pp. 171–212.

Third World Network (1989) 'Toxic waste dumping in the Third World', *Race and Class*, 30,3, pp. 47–57.

Thomas, A. and Allen, T. (2001) 'Agencies of development', in Allen, T. and Thomas, A. (eds) *Poverty and Development into the Twenty-first Century*, Oxford University Press, Oxford, pp. 189–218.

Thomas, J. and Evans, J. (2010) 'There's more to life than GDP but how can we measure it?', *Economic and Labour Market Review*, 4,9, pp. 29–36.

Thompson, J., Millstone, E., Scoones, I., Ely, A., Marshall, F., Shah, E. and Stagl, S. (2007) *Agri-food System Dynamics: Pathways to Sustainability in an Era of Uncertainty*, STEPS working paper 4, STEPS Centre, Brighton.

Todaro, M.P. (1997) *Economic Development in the Third World*, seventh edition, Longman, London.

Turner, R.K. (1988) *Sustainable Environmental Management*, Belhaven, London.

UNCTAD (2004) *The Least Developed Countries Report: Linking International Trade with Poverty Reduction*, United Nations, New York and Geneva.

UNCTAD (2011) *World Investment Report*, United Nations, New York.

UN-DESA (2010) *The MDG Report, 2010*, UN Department of Economic and Social Affairs, New York.

UNDP (2002) *Human Development Report*, Oxford University Press, Oxford.

UNDP (2003) *Making Global Trade Work for People*, Earthscan, London.

UNDP (2006) *Human Development Report 2006: Beyond Scarcity: Power, Poverty and the Global Water Crisis*, UNDP, New York.

UNDP (2007) *Human Development Report 2007/8: Fighting Climate Change: Human Solidarity in a Divided World*, UNDP, New York.

UNDP (2009) *Human Development Report 2009: Overcoming Barriers: Human Mobility and Development*, UNDP, New York.

UNDP (2010) *Human Development Report 2010: The Real Wealth of Nations: Pathways to Human Development*, UNDP, New York.

UNEP (2007) *Geo-4 Global Environment Outlook Environment for Development*, UNEP, Nairobi, Kenya.

UNEP (2008) *Vital Water Graphics – An Overview of the State of the World's Fresh and Marine Waters*, second edition, UNEP, Nairobi, Kenya.

UNEP (2011) *Environmental Assessment of Ogoniland*, UNEP, Nairobi, Kenya.

UNCHS (United Nations Centre for Human Settlements) (1996) *An Urbanising World: Global Report on Human Settlements*, Oxford University Press, Oxford.

UNCHS (2001) *Cities in a Globalising World: Global Report on Human Settlements, 2001*, United Nations Human Settlements Programme, Earthscan, London.

UNHSP (United Nations Centre for Human Settlements) (2006) *State of the World's Cities 2006/7*, United Nations Human Settlements Programme, Earthscan, London.

UNHSP (2008) *Ethiopia: Addis Ababa Urban profile*, UNHSP, Nairobi, Kenya.

UNHSP (2009) *Planning Sustainable Cities: Policy Directions, Global Report on Human Settlements*, abridged edition, Earthscan, London.

UNHSP (2010) *State of the Worlds Cities 2010/11, Bridging the Urban Divide*, Earthscan, London.

UNICEF (2009) *The State of the World's Children: Executive Summary*, UNICEF, New York.

UNICEF (2010) *Levels and Trends in Child Mortality*, UN Inter-agency Group for Child Mortality Estimation, UNICEF, New York.

United Nations (2000) *United Nations Millennium Declaration*, United Nations, New York.

United Nations (2010) *The Millenium Development Goals*, United Nations, New York.

UNPD (United Nations Population Division) (2009) *World Population Prospects: The 2008 Revision*, United Nations, New York.

Unwin, T. (ed.) (2009) *ICT4D: Information and Communication Technology for Development*, Cambridge University Press, Cambridge.

Verkaik, R. (2010) 'Trafigura faces £105m legal bill over dumping of toxic waste', *Independent*, 11 May.

Vidal, J. (1998) 'Woman power halts work on disputed Indian dam', *Guardian*, 13 January.

Vidal, J. (2008) 'Change in farming can feed world', *Guardian* 16 April.

Vidal, J. (2010) 'How food and water are driving a 21st-century African land grab', *Observer*, 7 March.

Vodafone (2009) *India: The Impact of Mobile Phones*, Vodafone Policy Paper Series 9, available online at http:/www.vodafone.com/ (accessed 6 February 2012).

Vorley, B. and Berdegue, J. (2001) *The Chains of Agriculture*, IIED, London.

Wackernagel, M. (1998) 'The ecological footprint of Santiago de Chile', *Local Environment*, 3,1, pp. 7–25.

Walker, G., Fairburn, J., Smith, G. and Mitchell, G. (2004) *Environmental Quality and Social Deprivation*, R&D Technical Report E2–067/1/TR, available online at www.environment-agency.gov.uk/commondata (accessed 30 January 2004).

Walker, G.W., Mitchell, G., Fairburn, J. and Smith, G. (2005) 'Industrial pollution and social deprivation: evidence and complexity in evaluating and responding to environmental inequality', *Local Environment*, 10,4, pp. 361–77.

Waraich, O. (2010) 'Record rains but Pakistan is dying for water', *Independent on Sunday*, 1 August, pp. 33–4.

Watts, M.J. (1994) 'Living under contract: the social impacts of contract farming in West Africa', *Ecologist*, 24,4, pp. 130–4.

WCED (World Commission on Environment and Development) (1987) *Our Common Future*, Oxford University Press, Oxford.

Webb, T. (2011) '"Let down by the West", the world's poor wait for arrival of electricity', *The Times*, 18 July.

Werksman, J. (1995) 'Greening Bretton Woods', in Kirkby, J., O'Keefe, P. and Timberlake, L. (eds) *The Earthscan Reader in Sustainable Development*, Earthscan, London, pp. 274–87.

Werksman, J. (ed.) (1996) *Greening International Institutions*, Earthscan, London.

White, H. (2008) 'The measurement of poverty', in Desai, V. and Potter, R.B. (eds) *The Companion to Development Studies*, second edition, Hodder Education, London, pp. 25–30.

WHO (World Health Organisation) (2000) *Global Water Supply and Sanitation Assessment 2000 Report*, WHO, Geneva.

WHO (2011) *World Health Statistics*, WHO, Geneva.

Wilkinson, R. and Picket, K. (2009) *The Spirit Level – Why More Equal Societies Almost Always Do Better*, Penguin, London.

Williams, A. and MacGinty, R. (2009) *Conflict and Development*, Routledge, Abingdon.

Williams, G., Meth, P. and Willis, K. (2009) *Geographies of Developing Areas: The Global South in a Changing World*, Abingdon, Routledge.

Willis, K. (2011) *Theories and Practices of Development*, second edition, Routledge, Abingdon.

Willett, S. (2001) 'Introduction: globalisation and insecurity', *IDS Bulletin*, 32,2, pp. 1–12.

Woodhouse, P. (2009) 'Technology, environment and the productivity problem in African agriculture: comment on the World Development Report 2008', *Journal of Agrarian Change*, 9,2, pp. 263–76.

Wolford, W. (2008) 'Land reform in the time of neo-liberalism', in Mansfield, B. (ed.) *Privatization: Property and the Remaking of Nature–Society Relations*, Blackwell, Oxford, pp. 156–75.

Wolsink, M. (2007) 'Wind power implementation: the nature of public attitudes: equity and fairness instead of "backyard motives"', *Renewable and Sustainable Energy Reviews*, 11,6, pp.1188–207.

World Bank (1990) *World Development Report*, Oxford University Press, Oxford.

World Bank (1992) *World Development Report*, Oxford University Press, Oxford.

World Bank (1993) *The East Asian Miracle*, Oxford University Press, Oxford.

World Bank (1994) *World Bank and the Environment, Fiscal 1993*, World Bank, Washington.

World Bank (2001a) *Attacking Poverty, World Development Report 2000/01*, World Bank, Washington.

World Bank (2001b) *Making Sustainable Commitments: An Environment Strategy for the World Bank, Summary December 2001*, World Bank, Washington.

World Bank (2003a) *World Development Report: Sustainable Development in a Dynamic World*, World Bank, Washington.

World Bank (2003b) *Annual Report, 2003*, World Bank, Washington.

World Bank (2006) *Kingdom of Morocco Poverty and Social Impact Analysis of the National Slum Upgrading Program*, Final Report No. 36545-MOR, IBRD, Washington.

World Bank (2008a) *Environmental Sustainability: An Evaluation of World Bank Group Support, Independent Evaluation Group*, IBRD, Washington.

World Bank (2008b) *World Development Report 2008: Agriculture for Development*, IBRD, Washington.

World Bank (2009a) *World Bank–Civil Society Engagement, Review of Fiscal Years 2007–2009*, World Bank, Washington.

World Bank (2009b) *Global Monitoring Report*, IBRD, Washington.

World Bank (2009c) *The World Bank Group Environment Strategy, Concept Note*, Environment Department, Washington.

World Bank (2010a) *World Development Report: Development and Climate Change*, World Bank, Washington.

World Bank (2010b) *World Development Indicators*, World Bank, Washington.

World Bank (2010c) *Annual Report 2010*, IBRD, Washington.

WRI (World Resources Institute) (1992) *World Resources, 1992–93*, Oxford University Press, Oxford.

WRI (1994) *World Resources, 1994–95*, Oxford University Press, Oxford.

WRI (1996) *World Resources, 1996–97: The Urban Environment*, Oxford University Press, Oxford.

WRI (2003) *World Resources, 2002–04*, Oxford University Press, Oxford.

Worth, J (2010a) 'Blood on the Summit floor', *New Internationalist*, 429, pp. 22–3.

Worth, J. (2010b) 'Taking on Tarmageddon', *New Internationalist*, 431, pp. 4–7.

Young, E. (2011) *Food and Development*, second edition, Routledge, Abingdon.

Zobel, G. (2009) 'We are millions', *New Internationalist*, December, pp. 21–4.

Index

Page numbers in *italics* denote a table/figure.